M54

AF725606

Developing Rates for Small Systems

Second Edition

American Water Works Association

Manual of Water Supply Practices—M54–Edition 2

Developing Rates for Small Systems

Managing Editor: Melissa Valentine
Production: Sandra Lankenau, Melanie Yamamoto (Cover)

If you find errors in this manual, please email books@awwa.org. Possible errata will be posted at www.awwa.org/resources-tools/resource.development.groups/manuals-program.aspx.

Library of Congress Cataloging-in-Publication Data

Library of Congress Cataloging-in-Publication Data

Names: Bradley, Daniel T., author. | Giardina, Richard D., author. | Matthews, Paul L., author.
Title: M54 : developing rates for small systems / by Daniel T. Bradley, Richard D. Giardina, Paul L. Matthews.
Other titles: Manual 54
Description: Second Edition. | Denver, CO : American Water Works Association, [2017] | Includes bibliographical references and index.
Identifiers: LCCN 2016048765 | ISBN 9781625760159 (alk. paper)
Subjects: LCSH: Water utilities--Rates. | Water utilities--Management.
Classification: LCC HD4456 .B73 2017 | DDC 363.6/10688--dc23 LC record available at https://lccn.loc.gov/2016048765

ISBN-13 978-1-62576-015-9
eISBN-13 978-1-61300-276-6

Printed in the United States of America
American Water Works Association
6666 West Quincy Avenue
Denver, CO 80235-3098
awwa.org

Contents

Figures

This page intentionally blank.

Tables

This page intentionally blank.

Preface

PURPOSE

In 2012, the American Water Works Association (AWWA) published the sixth edition of AWWA Manual M1, *Principles of Water Rates, Fees, and Charges*. That manual provides comprehensive guidance on establishing rates and charges for water service. The methods and examples contained in Manual M1, while recommended and appropriate for many water utilities, may prove to be overly complex and burdensome for others. This burden applies particularly to the development of cost-of-service rates. Such rates require the determination of revenue requirements by customer class before designing and adopting a rate structure.

Recognizing the limitations Manual M1 may have for smaller systems, or larger systems without sufficient data or rate-making capability, the AWWA Rates and Charges Committee (authors of Manual M1) developed the idea for a "Manual M1-type" alternative. A rate manual for small systems provides guidelines for the development of rates for utilities that lack the data and resources needed to apply the methods described in Manual M1. For these utilities, this manual—*Developing Rates for Small Systems*—provides guidance in developing rates when data and information may be lacking. The Rates and Charges Committee encourages all utilities to use Manual M1 as the primary guidance tool but believes this manual fills a need for small systems.

INTENDED AUDIENCE

This manual is intended to serve as a resource for managers (and others) of small water systems in the determination of rates for water service. Other resources, for example, Manual M1 and state and local government agencies, may also be of value in the preparation of rates for small systems. However, this manual focuses on the unique attributes of small systems as related to financial planning and rate design. Managers and operators of small systems and their advisors (financial, accounting, and engineering professionals) should find this manual useful in preparing water rates that recover costs using generally accepted cost-based practices.

For purposes of this manual and for those who might benefit from its use, there is no reason to specifically define a small system in terms of connections or population. Rather than define a small system in either of these terms, and therefore restrict or limit the applicability of this manual, the characteristics of the systems to which this manual may apply are the primary consideration. For example, such factors include

- A lack of or limited customer or demand data
- A system that predominately serves only a single class of customers, e.g., residential accounts
- Limited system peak-day and peak-hour data
- Limited or nonexistent fixed-asset data

Regardless of size, water systems with these and other similar characteristics will likely benefit from this manual. So although this manual is generally written for systems serving a limited population, any system having one or more of the characteristics listed above may also find it useful.

At the same time, this manual may not be applicable for many water systems. Systems with a rapidly growing or diverse customer base, those with a large industrial customer or customer base with significant outside city or wholesale service arrangements, or those with contracts requiring cost-of-service–based rates should refer to Manual M1 for guidance in financial planning, cost allocation methods, and rate design. Systems seeking to materially modify their rate design approach would also benefit from the information contained in Manual M1. In addition, Manual M29, *Water Utility Capital Financing*, should be reviewed as well when developing a financial plan.

UNDERLYING FINANCE AND RATEMAKING PRINCIPLES

The AWWA has established policies for finance, accounting, and rates. These AWWA policies provide direction to AWWA members and the volunteers that produced this manual. These policies are updated and generally available from AWWA. The policies included in the preparation of this manual were those revised by the Board of Directors on January 17, 2010.

The relevant AWWA policies for this manual include:

1. Water utilities' revenues from water service charges, user rates, and capital charges (e.g., impact fees and system development charges) should be sufficient to enable utilities to provide for:

    - Annual operation and maintenance expenses
    - Capital costs (e.g., debt service and other capital outlays)
    - Adequate working capital and required reserves

2. Water utilities should account for and maintain their funds in separate accounts from other governmental or owning entity operations. Water utility funds should not be diverted to uses unrelated to water utility services. Reasonable taxes, payments in lieu of taxes, and/or payments for services rendered to the water utility by a local government or other divisions of the owning entity may be included in the water utility's revenue requirements after taking into account the contribution for fire protection and other services furnished by the utility to the local government or to other divisions of the owning entity.
3. Water utilities should adopt a uniform system of accounts based on generally accepted accounting principles. Utility practices should generally follow the accounting procedures outlined in AWWA's accounting textbook, *Financial Management for Water Utilities: Principles of Finance, Accounting, and Management Controls.*
4. Modifications may be made to satisfy the financial and management control reporting needs of the utility and to meet the requirements of legislative, judicial, or regulatory bodies.
5. Water rate schedules should be designed to distribute the cost of water service equitably among each type and class of service. Rate-setting practices that are not based on cost of service may be appropriate in some situations, subject to legal review and approval, provided they reflect market conditions, the benefits received by the users of the service, and an appropriate balance of the goals and objectives essential to the public good. Any rate-setting practice not based on cost of service that is implemented by a utility should be fully disclosed to its customers, regulators, and the financial community. Such disclosure should identify each such rate-setting practice, its expected benefit, and its impact on the utility's customers.
6. Water utilities should maintain asset records that detail sufficient information to provide for the monitoring and management of the physical condition of infrastructure. These asset records should also support planned and preventive maintenance programs and budgets adequate to maintain the utility's assets at a level of service consistent with good utility practice. Utilities should annually provide comparative information to customers, the financial community, and

the general public about the utility's sustained capability to provide water service and generate revenue levels necessary to protect the financial investment of others in the utility. Such information could include historical expenditures for renewal and replacement during each of the past several years, as well as the revenues that would be generated under planned and adopted rates to support renewal and replacement during each of the next several years

OVERVIEW OF THE MANUAL

This manual contains five chapters and three appendices:

- Chapter 1: Basics of Water Ratemaking—presents an overview of the rate-setting process.
- Chapter 2: Customer Account and Usage Data—discusses the importance and use of system and customer account and usage data in the rates-setting process and begins a case study that presents this type of data for an example small utility.
- Chapter 3: Preparing a Financial Plan and Evaluating Revenue Sufficiency—discusses the types of costs often encountered in a small water utility, the establishment of reserves, and the process of developing a projection of costs and the compilation of a cash flow statement for use in projecting the need for future revenue adjustments. Continuing the case study, this chapter calculates revenues from current rates based on the customer and usage data contained in chapter 2. This chapter also discusses the process of determining revenue requirements and presents alternatives (as part of the case study) for increasing revenues, with consideration given to the use of reserves and the impact on customers.
- Chapter 4: Rate Design—discusses considerations and alternatives in the design of rates and illustrates a rate-design option based on the case study developed in the previous chapters.
- Chapter 5: The Rate Adoption Process—presents a variety of issues associated with the rate design process, including system development charges and funding of capital infrastructure, rates by customer class, affordability, public involvement, and the regulatory process for approval of rates (for investor-owned and some government-owned utilities).
- Appendix A: Alternative Cash Flow Plans—presents complete alternative cash flow plans.
- Appendix B: Examples of Financial Policies—presents various financial policies that utilities should consider.
- Appendix C: Frequently Used Financing Sources for Water Utilities—presents possible funding alternatives for small water utilities.

This page intentionally blank.

Acknowledgments

The AWWA Management and Leadership Division gratefully acknowledges the contributions made by those volunteers who drafted, edited, and provided the significant and critical commentary essential to updating M54. The Editorial Review Board members dedicated countless hours in the final stages of preparation of this edition to ensure the overall technical quality, consistency, and accuracy of the manual.

Editorial Review Board Members

Daniel T. Bradley, Oak Lodge Water District, Portland, Ore.

Richard D. Giardina, Raftelis Financial Consultants Inc., Centennial, Colo.

Paul L. Matthews, Tualatin Valley Water District, Beaverton, Ore.

Margaret Medellin, American Water Works Association, Denver, Colo.

William B. Zieburtz Jr., Black & Veatch, Alpharetta, Ga.

Chris Gonzales, FCS Group, Redmond, Wash.

Contributors to the Second Edition

Joy M. Barrett, Rural Community Assistance Partnership, Boulder, Colo.

Richard Fedder, Woodard & Curran, Andover, Mass.

William L. Golightly Jr., Mount Pleasant Water Works, Mount Pleasant, S.C.

Brian Jewett, Black & Veatch, Temecula, Calif.

Shawn W. Koorn, HDR Engineering, Bellevue, Wash.

Daniel D. Lanning, 3rd Generation Engineering, Dallas, Texas

Bryan A. Mantz, Public Resources Management Group Inc., Maitland, Fla.

John M. Mastracchio, Red Oak Consulting, An ARCADIS Group, Clifton Park, N.Y.

Paul L. Matthews, Tualatin Valley Water District, Beaverton, Ore.

Lawrence A. McCartney, City of Burleson, Burleson, Texas

Kathryn Sorensen, City of Phoenix, Mesa, Ariz.

Ronald W. Thomas, City of Dahlonega, Dahlonega, Ga.

Alexis F. Warmath, Raftelis Financial Consultants Inc., Charlotte, N.C.

Christopher P. Woodcock, Woodcock & Associates Inc., Northborough, Mass.

This page intentionally blank.

Chapter 1

Basics of Water Ratemaking

BASIC PREMISE

Water utility services are best provided by autonomous and financially independent utilities. Autonomous service means that the utility does not rely in any substantial way on outside agencies or entities for financial support. This also implies that the revenues collected by the utility are sufficient to operate and maintain the fixed assets as well as meet the ongoing operational needs of the utility over time while minimizing long-term financial costs for the users.

WATER UTILITY SERVICE

Although water itself, whether obtained from surface water or groundwater sources, is often free, the collection, treatment, and distribution of water to customers are not free. Services that are typically provided by water utilities include: water supply, storage, treatment, transmission, distribution, fire protection, metering, administration, billing, and customer service. The cost of each of these functions may be tracked in the utility's cost accounting system by various levels of detail. The cost of water utility service is typically expressed in annual operations and maintenance and annual capital costs.

FINANCIAL SELF SUFFICIENCY

The financing of water utilities in North America relies substantially on rates and fees charged to the utility's customers. Other sources of financing (e.g., property tax revenue, support from the community's general governmental funds, federal and state grants) have become increasingly less common. Meeting the financial needs of the utility has become an important responsibility of the utility's management and policymakers.

THE FINANCIAL PLAN

A financial plan is (see chapter 3) used by a utility to ensure its long-term capital and operating needs are met by future water rates. Figure 1-1 presents a basic schematic of a financial plan. In addition to annual operation and maintenance costs the plan should also address the capital costs needs of the utility. For utility planning, a multiyear financial plan is recommended that determines the annual revenue needs to prudently fund the utility's operating and capital needs.

An engineering master plan is the ideal starting point for determining capital costs. An engineering study of the infrastructure needed to serve projected levels of water demand in the longer term provides such information. The next step is the development of a capital improvements program (CIP). A CIP is a systematic listing of needed improvements ordered by priority and year. It usually spans a minimum of five years but may be longer. Ongoing operational costs also need to be studied. These include treatment costs, staff, administration, ongoing maintenance, meter reading, billing, and so on.

COST-OF-SERVICE RATE STUDY

Figure 1-2 presents six commonly followed steps to conduct a cost-of-service rate study. These steps are more fully described in AWWA Manual M1. A brief description of the process follows.

The first step in developing cost-of-service rates is to determine the revenue requirements. The revenue requirements for most utilities are taken directly from the utility's financial plan discussed in later in chapter 3 of this manual. The rates calculated by the cost-of-service study are set to generate an amount of revenue equal to the revenue requirements.

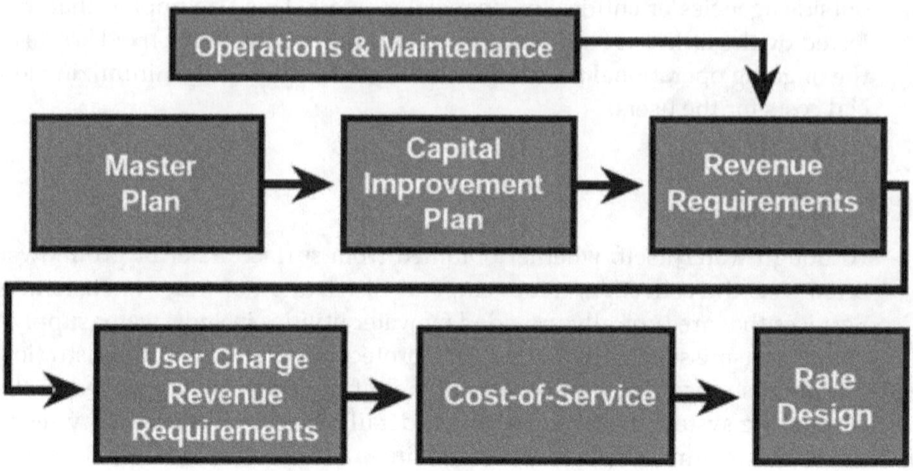

Figure 1-1 Schematic of a financial plan

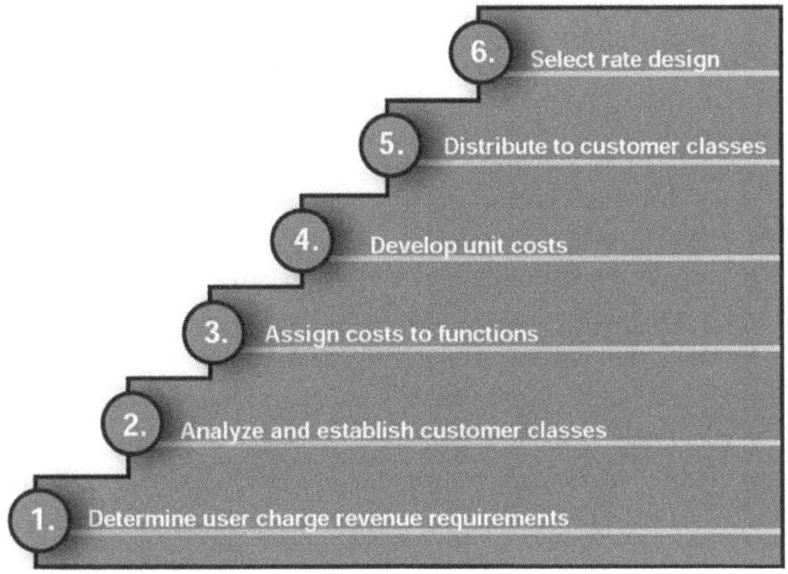

Figure 1-2 Cost-of-service steps

As a next step, in addition to determining revenue requirements, a cost-of-service study will often examine how the utility places customers into classes. Some small systems may have only one class of customers. Others may have separate classes for single-family residential, multifamily residential, commercial, institutional, irrigation, and so on. When a utility separates customers into classes, it generally adopts separate rates for each class to recognize the differences in the costs incurred to serve each class.

The third step in a cost-of-service study is normally to assign costs to functions. Common water system functions include source of supply, treatment, pumping, storage, transmission and distribution, meters and services, customer service, and billing.

After assigning costs to water system functions, the rate analyst develops unit costs for each function based on how the utility's customers use the system. These unit costs serve as the basis for the cost-of-service rates.

Once the unit costs by function are determined for each function, the analyst distributes these costs to each customer class by multiplying the units of service demanded by each customer class by the unit costs for each function. The product resulting from this multiplication is called the *cost-of-service for the customer class.*

The final step in a typical cost-of-service study is to select a rate design. The rate design is the specific set of rates for each class of customer to recover the cost-of-service for that customer class. Most rate designs use the unit costs developed in a cost-of-service study to set the fixed charges (i.e., that part of the rate design that does not vary with the amount of water used but may vary with the size of the meter) and the volume charges. Chapter 4 of this manual discusses the rate design process.

IMPLEMENTING RATES

After a rate design has been selected and rates calculated, the utility should consider engaging its customers in a public process. That process could range from a simple public presentation to more complicated presentations using study/focus groups with representative customers. Chapter 5 discusses the rate adoption process in more detail.

PERIODIC REVIEWS

Regular reviews of the utility's financial status are critically important to stay abreast of revenue needs, meet operational needs, and help the utility to achieve long-term cost efficiencies. Many utilities benefit from conducting these reviews annually as part of their budgeting process. In those cases, the utility may update its financial plan based on changes that have occurred since adopting the current rates. Minimally, the periodic review should include:

- Assessments of revenue received versus expenses
- Operational cost trends versus budgets
- Capital cost expenditures trends versus capital plan
- Maintaining sufficient operational and capital reserves
- Condition assessment of major fixed assets
- Trends in customer use (including the effects of conservation)

Chapter 2

Customer Account and Usage Data

This chapter discusses a utility's customer base, the specific customer information a water utility should consider in order to calculate water rates and charges, and how that information should be applied. This chapter also introduces the example utility that will serve as the case study throughout this manual.

Appropriate water rate setting should embrace the concepts of fairness, equity, and cost causation.[1] These concepts are applicable to both the utility's customers and to the utility itself. First, water rates should be structured so that each class of customer and, to the extent practicable, the individual customers within each class are paying their appropriate share of the cost incurred by the utility in providing service to those customers or classes. Also, the utility should try to avoid situations in which one customer or customer class is unnecessarily subsidizing another. In accomplishing these rate-setting objectives, it is necessary to ensure that the rates are based on accurate customer information and data.

Second, the total revenue that is collected by the water utility should be reflective of the cost of operating the utility. To meet this objective, a utility should collect revenues sufficient to meet its revenue requirements and satisfy covenants associated with its outstanding indebtedness and other legal or self-imposed financial performance measures (e.g., financial performance targets recognized as prudent by the utility industry and credit rating agencies; see chapter 3).

[1] "Cost causation" is the principle of attributing or allocating costs to specific customers based on whether the customers "caused" the costs either because (1) the costs are brought into existence as a direct result of providing service to the customers; or (2) the costs are avoided if the service to the customers is not provided.

A utility should also have a stable rate revenue stream over the long run to enhance its overall financial integrity. For the water utility to successfully meet all of the aforementioned goals when setting water rates, it must have a good foundation of customer account and water usage data on which to establish those rates. At a minimum, in order to promote fairness and equity in the rates, the utility must know the number of customers being served and the amount of water being consumed within a given year or billing cycle. The utility must also know the overall revenue requirements to effectively operate the water utility (see chapter 3). Additional customer information may serve to improve the utility's ability to enhance the fairness and equity of its rates and charges, e.g., water use by class and meter size.

Accurate customer data are critical for establishing supportable water rates that are fair and equitable to customers and that will generate sufficient revenue for the utility. Further, good records are necessary to adequately assess the impact of alternative rate structures on issues such as revenue sufficiency and stability, individual customer or customer class equitability, and conservation goals and objectives. By far the most significant source of revenue for most water utilities is that produced from water sales. The application of a well-designed rate structure to the number of bills and/or metered water usage produces the revenues needed to sustain the utility's financial well-being.

CUSTOMER INFORMATION

The primary source of customer information is the utility's customer billing system. Much of the information that is necessary to produce a utility bill is the same information that is required to establish water rates. The development of a rate structure can be hampered by inaccurate customer billing records, resulting in rates that do not generate sufficient revenue to meet the utility's revenue requirements, satisfy its debt covenants, and provide fair and equitable charges to its customers. In the absence of adequate or accurate customer records, the utility should make an investment to improve its customer information before moving forward with a comprehensive rate review. Prior to such an investment, however, the utility should attempt to identify various shortfalls and problems with its existing customer records. Possible improvements could include a customer profile survey, the re-creation of records, and compilation of other information that could assist in the development of equitable rates and charges (e.g., determining equivalent residential units [ERUs] as discussed later in this chapter) dependent on the particular shortfalls in the utility's customer database.

It is also critical to understand the limitations of the existing billing system in implementing alternative rate structures. Rate structures that require information that is not currently maintained in the existing database or that cannot be collected and entered economically or efficiently should not be considered unless plans exist to modify the billing system.

As previously stated, it is likely that much of the information necessary to develop a schedule of water rates is contained and available within the utility's billing system. However, accessing that data in a timely and cost-effective fashion may not be possible. It is important that, as part of the rate-study process, ample time is allowed for the identification, extraction, compilation, and review of the customer data from the electronic database. There may be technical challenges in each step of this process that slow access to this information. These technical issues should be identified and corrected if possible. Once retrieved from the database, this information should be reviewed and audited for accuracy and reasonableness. This data will be the foundation of the utility's water rates, and errors in the data could result in inappropriate rates or in the utility collecting too little or too much revenue.

IMPORTANCE OF CUSTOMER RECORDS

For the purposes of water rate setting, accurate and detailed customer records are important in the areas of (1) number of customers, (2) metered consumption, (3) billed revenue data, (4) peak-period demand data, and (5) information on public and private fire services. For some utilities, not all of this information is readily accessible or even available. The example illustrated in the balance of this manual is based on the following data:

- Number of customers
- Metered consumption
- Billed revenue

Number of Customers

As a starting point, most utilities maintain certain standard customer-related statistics. This information typically includes a listing of water utility customers and their billing addresses. Along with this information, most utilities will have details related to the size of each customer's meter and, if multiple rate classes are used, the class for each customer. This type of information does not generally change between billing periods and forms the basis of the customer record. Along with the basic information are other data that generally are updated for each billing period. These include metered consumption data, payment history, and service call history. Data for past billing periods may be available on a monthly/quarterly/semiannual or aggregated year-end basis, depending on the sophistication of the utility billing system and the frequency of billing.

A utility should maintain records regarding the number of water bills it has issued in terms of meter size and customer class. This detailed information is necessary to accurately develop customer-related charges such as a minimum charge (which may include some water usage allowance) or service charge (billing or meter units that do not include a water usage allowance). These base or fixed charges are common to almost every type of alternative rate form and may be significant in terms of providing the water utility a stable source of revenue.

Table 2-1 shows an example tabulation of the number of metered accounts by meter size and by customer class, as well as the number of unmetered accounts by class. As discussed below, this information can represent a year-end accounting, annual average, monthly average, or other billing period. The basis of reporting should be clearly specified in the database so the reporting period is clear to the user.

Within any given water utility, it is common to bill all customers on the same frequency, although billing frequencies may vary by classification of customer. For example, residential accounts may be billed quarterly while other accounts are billed monthly. This variation may be done for a variety of reasons, but the result is that there may not be a one-to-one correspondence between the number of customers within a class and the number of water bills generated. For the example illustrated in this manual, all customers are billed monthly. In Table 2-2 all customers (see Table 2-1) are assumed to have been active and receiving service for each month of the year; as such, the number of accounts multiplied by 12 yields the number of bills shown in Table 2-2. In reality this is rarely the case because customers open and close accounts as they move in and out of a community.

Calculating number of customers. The number of customers can be calculated in a variety of ways, depending on the amount of detail available. It is most common to assume that the number of customers is equal to the number of metered and unmetered services.

End of year. The end-of-year customer count may be used to determine the number of customers. An advantage with this count is that it reflects the net number of accounts added throughout the year and, in the absence of a known change in the customer base, forms a reasonable basis for projecting accounts for the coming years. For a smaller utility

Table 2-1 Number of customers by meter size for the example utility

	Residential	Nonresidential	Total
Unmetered Accounts	200	40	240
Metered Accounts			
⅝ in. and ¾ in.	1,900	80	1,980
1 in.	—	15	15
1 ½ in.	—	5	5
2 in.	—	4	4
3 in.	—	3	3
4 in.	—	2	2
6 in.	—	1	1
Total metered accounts	1,900	110	2,010
Total all accounts	2,100	150	2,250

Table 2-2 Number of annual water bills sent to customers by meter size

	Residential	Nonresidential	Total
Unmetered Accounts	2,400	480	2,880
Metered Accounts			
⅝ in. and ¾ in.	22,800	960	23,760
1 in.	—	180	180
1 ½ in.	—	60	60
2 in.	—	48	48
3 in.	—	36	36
4 in.	—	24	24
6 in.	—	12	12
Total metered accounts	22,800	1,320	24,120
Total all accounts	25,200	1,800	27,000

that does not experience significant turnover in accounts, this method may be adequate for projecting the number of future customers for rate-setting purposes. In a steadily growing system, if the end-of-year values are used to estimate revenues for the year in question, these values may overstate the revenue.

Annual average. The annual-average customer count is the average of the customer count from the beginning of the year and the end of the year and, as such, corresponds more closely with the costs incurred to produce water for that same period of time. However, it has the disadvantage of understating the customer count in a growing utility and overstating the customer count in a utility with a decreasing customer base. The practical risk is that if the average is used for rate setting, the actual revenues received by the utility may not be the same as the forecasted revenues. Care should also be taken if this approach is used by a utility with significant seasonal variations in the number of accounts, i.e., resort and "second home" communities.

Monthly average. The monthly average customer count is calculated by taking the arithmetic average of the 12-month end-customer counts to produce the average number of customers per year. This figure can be more accurate than either the end-of-year or annual-average values because it captures the month-to-month changes in customers that can occur. This approach will produce a more accurate per-customer cost, but it also shares the risk of misstating forecasted revenues if the annual average is quite different from the year-end customer count balance. If the utility has a significant seasonal population, this approach may be more accurate.

Metered Consumption

In order to design and evaluate alternative rate structures and to determine those that best meet the utility's goals, it is important to have detailed customer consumption records. Available information may be limited to metered volume by customer account for utilities that do not have multiple customer classes or use tiered rates (rates with two or more unit-volume rates and associated usage blocks or steps) and/or seasonal rates. At a minimum, individual customer data should be maintained on an annual basis, but preferably it should be maintained on a billing cycle basis (monthly, bimonthly, quarterly, etc.) for at least one full year.

Metered consumption is of particular importance in areas where weather can vary significantly from year to year, water usage is sensitive to weather patterns, and the utility currently has or is considering a rate structure that reflects this variability of demand. For example, in regions where discretionary uses such as lawn irrigation constitute a large proportion of annual or at least seasonal water sales, the utility may want to institute a water conservation rate structure to encourage efficient use of the system and resource. In such cases, water consumption by customer class by billing period is necessary to establish such a rate structure.

In the illustrative example developed for purposes of this manual, annual metered water consumption data were assumed to be readily available only for total residential and total nonresidential customers (metered). Water consumption information was not available by meter size. To determine the consumption for accounts with meters larger than 3/4 in., the utility manually aggregated the water bills to calculate consumption on the 30 accounts that comprise the 1-in. to 6-in. meter customers. An estimate was used to determine the water use for the unmetered nonresidential accounts based on the known consumption of the total metered accounts. Similarly, the unmetered residential account usage was based on the average usage per account for the metered residential accounts. This information is summarized in Table 2-3.

If possible, this type of billing data (ideally on a monthly basis or the utility's billing frequency) should be maintained for a period of 3 to 5 years in order to provide the

Table 2-3 Water use by meter size (1,000 gal)

	Residential (1,000 gal)	Nonresidential (1,000 gal)	Total (1,000 gal)
Unmetered Accounts	25,300	16,400	41,700
Metered Accounts			
⅝ in. and ¾ in.	240,000	7,500	247,500
1 in.	—	3,520	3,520
1 ½ in.	—	2,350	2,350
2 in.	—	3,000	3,000
3 in.	—	4,220	4,220
4 in.	—	7,510	7,510
6 in.	—	16,900	16,900
Total metered accounts	240,000	45,000	285,000
Total all accounts	265,300	61,400	326,700

utility with a database that encompasses a wide range of potential weather conditions and related water usage and demands on the system. The availability of this type of data enables the water utility to "test" various alternative rate structures against a variety of weather conditions. This information further allows for the analyses and establishment of necessary reserves or working capital balances that should be maintained to enhance the revenue stability of the utility during periods of lower-than-normal water usage caused by weather conditions or other events.

Another important reason to keep detailed records of metered consumption by class of customer and/or meter size on a billing period basis is to enable the development of a bill frequency distribution analysis, or bill tabulation analysis.[2] A bill tabulation analysis is particularly important in the development of block rate structures or in the establishment of lifeline rates, where a different unit rate is assigned to metered consumption that falls in a predetermined consumption or rate block. In these rate designs, it is necessary to know the percentage distribution of the annual water use by class in each of the consumption or rate blocks. Without that knowledge, there is a risk that the calculated water rate will result in revenues that may be significantly different from the utility's revenue requirements.

Billed Revenue

A third customer parameter (in addition to the number of customers and amount of metered consumption) that is important in the development of rates and that is maintained by most utilities is billed revenue. Most often this information is available by customer class and perhaps by meter size if the rate structure includes a charge that varies by meter size. This information is important for the utility to use as a benchmark when comparing actual billed revenue to the billed revenue anticipated from proposed new rates. In practice, this is done by applying existing rates to the customer account and consumption data to develop a "base" level of billings. The existing rates for the example utility are illustrated in Table 2-4 and are used to develop the revenues illustrated in Table 2-5.

The base billings can then be compared with actual billings. This comparison will allow the water utility a measure of the accuracy of the billing statistics or units of service with which it is working and will provide for a reasonable level of accuracy that the anticipated revenue will be achieved when the proposed rates are applied to these billing units. Table 2-5 shows an example of billings under existing rates based on the number of bills and metered consumption statistics developed in the previous tables. For the example utility, fixed revenues represent revenue from unmetered accounts and the fixed charge assessed all metered accounts. Volume revenues represent the volumetric or consumption portion of the customer's bill, i.e., the $1.50-per-1,000-gal rate applied to metered water use for the example utility. For the example, utility total revenues are calculated at $801,000 for the "base" year.

Table 2-4 Current rates

Type of Account	Rate
Unmetered Accounts	$25.00 per account per month
Metered Accounts	
Volume rate	$1.50 per 1,000 gal
Fixed charge	$12.50 per account per month

[2] This analysis is described and illustrated in AWWA Manual M1, appendix C.

Table 2-5 Current rate revenues by customer type and meter size

	Residential		Nonresidential		Total		
	Fixed	Variable	Fixed	Variable	Fixed	Variable	Combined
Unmetered Accounts	$60,000	$ –	$12,000	$ –	$72,000	$ –	$72,000
Metered Accounts							
⅝ in. and ¾ in.	285,000	360,000	12,000	11,250	297,000	371,250	668,250
1 in.	–	–	2,250	5,280	2,250	5,280	7,530
1 ½ in.	–	–	750	3,525	750	3,525	4,275
2 in.	–	–	600	4,500	600	4,500	5,100
3 in.	–	–	450	6,330	450	6,330	6,780
4 in.	–	–	300	11,265	300	11,265	11,565
6 in.	–	–	150	25,350	150	25,350	25,500
Total metered accounts	285,000	360,000	16,500	67,500	301,500	427,500	729,000
Total all accounts	$345,000	$360,000	$28,500	$67,500	$373,500	$427,500	$801,000

When a rate change is made, customer inquiries regarding their bills often increase. To assist staff in responding to such customer calls, it is helpful if the utility's billing database contains readily accessible historical billing information by individual customer. It is also helpful to use this information to establish water use profiles so that customer impacts can be estimated based on changes in rates. These profiles are good tools to have available when addressing and explaining rate adjustments or a new rate structure in public forums.

CUSTOMER CLASSES

Customer account and metered consumption information is used to establish customer classes for the purposes of water utility billing. Rates should be developed so as to assign cost responsibility to each individual customer served by the utility in a way that reflects the cost incurred by the utility in serving each customer. Often, it is neither economically practical nor possible to determine the cost responsibility and applicable rates for each individual customer served by a utility. However, it is common to determine the cost of providing service for classes of customers that have similar water use characteristics and also for special customers that have unusual water use or service requirements. One objective of effective ratemaking is to assign costs to classes of customers in such a manner that rates are equitable by approximating the cost of providing service to each customer class.

To achieve this objective, customers may be divided into classes that share common service characteristics and water use and demand patterns. An example of these characteristics can be found within the residential customer class. The residential customer class primarily consists of single-family or dual-family dwellings. This group typically uses the smallest water meters, commonly 5/8 in. or 3/4 in. In many utilities, lawn irrigation in the summer months by the residential customer class places a significant peak-demand requirement on the system. These characteristics, prevalent across most residential users, provide the common thread for creation of a residential customer class for purposes of aggregating customer data and analyzing costs.

Comparatively, other large water users may not rely on the same extensive network of smaller distribution mains for water service; these customers typically rely on larger water meters to measure their flow requirements. These customers often use water at a fairly constant rate throughout the day, resulting in a lower peak-demand requirement on the system when compared to the residential class. This distinction illustrates the importance of considering rates by customer class and the need to maintain usage data by customer class.

In very broad terms, the following are common definitions of customer classes:

Residential: One- and two-family detached dwelling units

Commercial: Multifamily apartment buildings and nonresidential, nonindustrial business enterprises

Industrial: Manufacturing and processing business enterprises

Wholesale: Master-metered customers that are generally located outside the jurisdictional limits of the water utility and that serve individual customers within their own service areas

As previously noted, one advantage of classifying customers into discrete classes or groupings is that it allows the utility to provide the starting point for considering different rates for each class of customer. This may be useful if the data exist to analyze the different demand characteristics and usage patterns of each class and, as appropriate, establish rates that maintain fairness among the different customer groups. Some utilities break down the four customer classes described above into more specific groups. For example, the commercial customer class may be separated into multifamily residential customers and commercial customers. Similarly, the industrial customer group may be subdivided into small industry, large industry, and special—the latter typified by a customer that is wholly unique to the utility and so significant that it is not reasonable to put that customer into a class with other customers. The special industrial customer could include a petroleum refinery, a pulp mill, or a power plant. Other classes may include consideration of different-sized commercial or municipal accounts.

For some utilities it may not be necessary to establish multiple classes of customers. Based on the principles of common service characteristics, water use and demand patterns, and jurisdictional limits, a small utility may identify only one, two, or perhaps three of these classes as being applicable to its particular circumstance. It is also less likely that a small utility serving a relatively homogeneous group of customers would develop specific groups within the classes. This would only occur if there were very unique and special circumstances for a particular customer or group of customers within that utility (e.g., a large university/college campus).

In some cases, data are not available to establish customer classes. In addition, there can be considerable variation in total use and demand characteristics within the traditional customer classes of residential, commercial, and industrial. For example, a bank with no public restrooms or lawn irrigation requirements will tend to have significantly lower consumption and different peaking patterns than a restaurant; yet both are typically classified as commercial. As an alternative, some utilities use meter size as a method to classify customers or to separate customers into subclasses.

In developing groupings of meter size for classification purposes, it is important to examine the total water use per customer in each class and, to the extent billing data are available, try to group customers that appear to have similar peaking characteristics. As an example, this latter analysis can be approximated by looking at use per customer during peak-demand periods (e.g., the summer) and use during nonpeak periods. An example of such a customer classification may be small class—5/8-in. and 3/4-in. meters; medium class—1-in. to 2-in. meters; large class—3-in. to 6-in. meters; and very large—larger than 6 in. For most smaller utilities, only two or three groupings should suffice.

Wholesale Customers

Wholesale customers can generally be thought of as other water suppliers that are sold water through a master meter at one or more points of delivery for resale to individual retail customers within the wholesale customer's service area. Usually, a wholesale customer is a separate municipality or water district adjacent to the supplying utility. The utility that supplies water to the wholesale customer typically has no responsibility for distribution of the water within the wholesaler's system and is not responsible for individual customer metering, billing, distribution system maintenance, or customer relations. The supplying utility typically sells treated water in bulk to another supplier. A utility providing wholesale service may benefit from the information and concepts discussed in AWWA Manual M1 with regard to the development of wholesale rates.

Unmetered Accounts

Accounts with unmetered water service present challenges in developing water rates. These challenges result from the difficulty in estimating the amount of water used by unmetered customers. Additionally, unmetered customers may use more water than similarly situated metered customers because they typically pay a flat charge regardless of water use. Therefore, these unmetered customers do not face a price signal to reduce water use. It is necessary to establish a separate class and rate for customers that have unmetered service. The utility may want to incorporate incentives in the water rates applicable to unmetered customers that would encourage such customers to change to a metered service, e.g., the water use levels assumed in developing the flat rate for unmetered customers are greater than the average use indicated for metered accounts or the utility may provide the meter and/or installation at no cost to the customers. Alternatively, the utility may establish a policy that requires all customers to be metered.

For the purposes of rate setting, it is necessary to establish some assumptions regarding the amount of usage and demand pattern for the unmetered customer. It is reasonable to make the simplifying assumption that customers who have unmetered services will use at least as much water as the same type of customer with a metered service and that the unmetered customer's water usage characteristics are similar to the same type of customer with a metered service. This usage level can then be incorporated into the rate analysis for purposes of developing a flat or fixed charge per account per billing period.

EQUIVALENT RESIDENTIAL UNITS

General

The fundamental purpose of an equivalent residential unit (ERU) is to "standardize" the customer base for rates and charges, given that a commercial account uses water differently than a single-family home. An ERU has multiple purposes including projecting revenue, allocating costs among customer classes, and calculating rates and charges. Growth-related fees such as system development charges (also known by other names such as *impact fees* or *capacity charges*) are typically charged on a per-ERU basis.[3] More information on system development charges is available in AWWA Manual M1. Also, the calculated ERUs for the utility's customer base may be used in developing rate structures other than the meter-based approach detailed in this manual.

---

[3] A system development charge (SDC) is a one-time charge assessed against new customers (applicants for service) or developers as a means of recovering all or part of the costs of existing or additional system capacity constructed for their use.

Utilities should ideally maintain records on the calculated number of equivalent residential units for each customer or applicant for utility service based on the adopted methodology for calculating such units. An ERU is a means of expressing different levels of customer demand for water in terms of an equivalent number of least-size residential demand accounts. One ERU is typically equal to the average daily demand of a ⅝-in. or ¾-in. individually metered residential account.

Different utilities may use alternative terms when referring to ERUs. Equivalent residential units are often called *equivalent residential connections* as well as *equivalent dwelling units* and a variety of other terms.

Most utilities have a specific level of service assigned to each ERU. If the utility has a level of service for a residential account of 300 gpd (for example), then other customer groups' accounts would be expressed in terms of the equivalent number of these least-size accounts. The total ERUs of a customer are a measure of the proportionate share of the utility's capacity that the customer is using.

ERU Calculation Methodologies

A utility should adopt a methodology for calculating the number of ERUs for all other customer types. Common methodologies for calculating the number of ERUs for nonresidential customers include basing the ERU count on

- Meter size
- Customer attributes
- Plumbing fixtures
- Historical flow

The meter size approach may be based on the rated maximum flow capacity for each meter size (i.e., based on the hydraulic capacity of the meter) in accordance with AWWA standards and as published in various AWWA manuals (e.g., Manuals M1 and M6). This method is generally the easiest method to apply and is common. The ERU value for each meter size reflects the meter's flow potential compared with that of the smallest-size residential meter. However, a shortcoming of the method is that it does not capture the differences in end-use characteristics of differing facilities with similar meter sizes. For example, an office building and a restaurant may have the same meter size and the same ERU count under the meter size approach, but the restaurant is likely to have a significantly higher water demand. Common ERU values under the meter size approach are shown in Table 2-6.

The customer attributes method is also a common method to determine the number of ERUs for a new customer. This method is based on the type of establishment and standardized usage characteristics rather than meter size or plumbing fixtures of the new structure. For example, the capacity demand of a restaurant may be based on the number of seats, while the capacity demand of an office building may be based on the square feet of the building. A major advantage to the customer attributes method is that it may be a better indicator of actual use compared to a meter-based or fixture-based approach that does not consider how the facility will be used. However, one downside is that compared to the meter-based or fixture-based approaches, the attributes method can be more administratively burdensome. There may be difficulty in surveying existing establishments, and some may not be easily categorized.

Some utilities determine the number of ERUs for a customer based on a count of the plumbing fixtures in the customer's facility. One advantage of this approach is that the number of ERUs is based on utility industry demand standards. Each fixture has the potential to handle a certain amount of flow, and the ERU value for the fixture reflects

Table 2-6 Common ERU values under the meter size ERU calculation approach

Meter Size	Meter Type	Rated Maximum Capacity (gpm)	Meter Equivalent / ERU Value
⅝ in.	Displacement	20	1.0
¾ in. (*)	Displacement	30	1.5
1 in.	Displacement	50	2.5
1 ½ in.	Displacement or turbine	100	5.0
2 in.	Displacement, compound or turbine	160	8.0
3 in.	Displacement	300	15.0
3 in.	Compound	320	16.0
3 in.	Turbine	350	17.5
4 in.	Displacement or compound	500	25.0
4 in.	Turbine	600	30.0
6 in.	Displacement or compound	1,000	50.0
6 in.	Turbine	1,250	62.5
8 in.	Compound	1,600	80.0
8 in.	Turbine	1,800	90.0
10 in.	Compound	2,300	115.0
10 in.	Turbine	2,900	145.0
12 in.	Turbine	4,300	215.0

* Utilities sometimes choose to assign an ERU value of 1.0 to both the ⅝-in. meter and the ¾-in. meter. If the ¾-in. meter is the smallest meter used by the utility, the utility may choose to calibrate the ERU values based on the ¾-in.–rated maximum capacity. In this case, the ¾-in. meter would have an ERU value of 1.0, a 1-in. meter would have an ERU value of 1.7 (50 gpm/30 gpm), a 2-in. meter would have a calculated ERU value of 5.3 (160 gpm/30 gpm), etc. For more detailed information about meter equivalents, reference AWWA Manual M1.

such potential. Another advantage of the plumbing fixture approach is that it recognizes the capacity needs for each individual customer—the fixture value links to the customer-specific demand requirements. The disadvantage of the fixture count approach is that it may or may not reflect differences in customer end-use characteristics. Moreover, there may be potential administrative difficulties in surveying existing establishments.

Under the historical flow approach, ERUs are determined based on historical actual usage data for the customer or for similar customers if the customer has not yet connected to the system. While this approach may be reasonable for approximating actual demand, the variance of annual usage over time may skew the calculation of demand. Moreover, historical records may not be available for customers that are comparable to a customer that has not yet connected to the system but for which the utility is attempting to calculate an ERU count. Finally, when attempting to calculate wastewater ERUs under this approach, water use as measured by a customer's meter may not necessarily correspond to wastewater demand due to outdoor/irrigation uses.

The ERUs for multifamily residential customers can be calculated based on any of the above methodologies. An equitable approach that is common for these types of customers is to base the calculation on the number of multifamily units behind the master meter. While some utilities consider a multifamily unit to be one ERU, such units are often assigned a value less than one ERU.

Historical usage statistics can help to determine an appropriate ERU factor for a mutifamily unit. For example, if the average monthly usage per unit in the utility system is 4,000 gal compared with 5,000 gal for a single-family residence, an ERU value of 0.8 (4,000 gal/5,000 gal) might be justified.

ERU calculation methodologies for many utilities can be found through accessing utility or local government websites or by performing searches on Municode (municode .com), American Legal Publishing Corporation (amlegal.com), or other websites containing libraries of local government code of ordinances.

Chapter 3

Preparing a Financial Plan and Evaluating Revenue Sufficiency

OVERVIEW OF THE FINANCIAL PLANNING PROCESS

The primary objective of the financial planning process is to ensure that the utility has the ability to obtain sufficient funds to develop, construct, operate, maintain, and manage its water system on a continuing basis, and in full compliance with federal, state, and local requirements. The financial planning process involves the development of a financial plan that evaluates the utility's anticipated revenue needs, defined by operating costs, capital costs, and established policy objectives. Common policy objectives that may impact revenue needs include renewal and replacement (R&R) funding, reserve levels, and debt service coverage among other factors.

The financial plan is an outcome of the financial planning process and is a critical tool to help maintain financial sustainability and establish creditworthiness to obtain favorable interest rates on borrowed funds. For utilities that are averse to debt or unable to issue it, a primary objective of the financial plan could be to avoid future debt issuance. An integral financial plan can be developed by extending the current year's budget projection (in summary or detailed form) over a multiyear period. Any known changes in the utility's operations during the projection period (e.g., expansion of the water treatment facilities or increases in the level of annual renewal/replacement needs) should be recognized in the financial plan projection. The financial plan is used as the framework within which the interrelationship of revenues, reserves, operating costs, and capital costs is analyzed. A study period of 3 to 10 years or more is generally used, depending on the plan's purpose. This period allows a determination of whether existing rates are adequate to cover costs (now and in future years), and if not, how much of an overall increase in rates is needed. A properly constructed financial plan operates like a flow-of-funds projection (e.g., cash inflows and outflows), consolidating the various projections of revenues and expenses into one document or analysis.

The first part of this chapter discusses approaches and procedures for developing a financial plan along with the utility-specific information and documentation that are needed to develop a multiyear plan. The case study example included in this chapter builds on the information presented in chapter 2 and shows the relationships between projected costs (capital, operation, and maintenance) and the revenue increases required to meet those costs, maintain prudent cash reserves, and comply with applicable revenue bond covenants.

The second part of this chapter describes the process of determining the pattern of annual revenue adjustments and establishing the test year revenue requirements for setting rates, which are directly related to the financial plan. The test year is the annual period for which rates are developed. Often, the test year corresponds to the upcoming budget year, but rates might also be developed for another year of the projection period and then adjusted to reflect differences in revenue requirements for prior or future years. For the case study example, rates are illustrated for a 5-year test period.

DEVELOPMENT OF THE FINANCIAL PLAN

The development of the financial plan involves the analysis and projection of revenues, expenses, cash flow, and key financial ratios. The plan results in an estimate of the annual cash expenditure needs of the utility and is the first step in the review of the sufficiency of existing rates. The objective of the financial plan is to determine the cash needs, or revenue requirements, and anticipated timing of utility costs in order to ensure that adequate funds are available to meet these requirements as they occur. The basic financial plan for a small government-owned utility normally includes projections of

- Revenues—operating and nonoperating
- Operation, maintenance, and administrative (OMAR or commonly O&M) expenses
- Debt service (principal and interest payments) on borrowed funds
- Capital costs
  - Allowances for normal annual repairs and replacement of existing facilities
  - Cash-financed expansions and improvements to the system
- Other sources and uses of cash (such as transfers out of the utility fund and the use of or contributions to reserves)
- Fiscal year beginning and ending cash balances
- Fiscal targets (such as reserves and coverage targets)

In addition to these financial plan elements, a utility may also have outstanding bonds or loans that have associated debt covenants that must also be considered in establishing the adequacy of the revenue under existing rates. These debt covenants may include a requirement that the utility maintain a minimum debt service coverage ratio. Generally, a debt service coverage ratio is the quotient of the utilities net revenue divided by a measure of annual debt service and is more fully described later in this chapter. The practical effect of the debt service coverage requirement is that utilities often need to generate more revenue than is strictly required to make their annual debt service requirement and cover O&M expenditures. The requirement is intended to protect the utility's credit quality and thereby the bondholders. The effect of these debt covenants is that a utility must maintain sufficient revenue to meet these requirements, and the financial plan must reflect these requirements.

A multiyear time horizon provides a long-term view for planning purposes and allows the utility to better anticipate and prepare for major financial impacts from changes in operations or capital requirements. For financial planning purposes, the number of years reflected in the financial plan should depend on the plan's purpose. While rate and bond feasibility studies typically consider a 5-year forecast period, a utility may want to develop a financial plan for a longer period of time to match the planning horizon reflected in the long-term master plan. The financial plan allows for the consideration of both the immediate financial impacts (rate impacts) of the current test year being reviewed, and enables the utility to plan for any future revenue requirement impacts during the planning period. The case study included in this chapter uses a 5-year period and is essentially composed of four interrelated tables. Table 3-1 is a 5-year projection of revenues based on the example utility's current rates. Table 3-2 separates O&M expenses in sufficient detail to allow independent estimates of significant cost categories. Table 3-3 includes the capital spending plan and the dedicated capital funding sources. Data from Tables 3-1, 3-2, and 3-3 are summarized in Table 3-4, which presents the "big picture" of cash flows and financial performance over the 5-year study period by incorporating the elements of anticipated operating revenue (from existing rates and other fees), cash balances, O&M expenses, and capital costs. In most instances, dollar values in these tables have been rounded to the nearest thousand.

Revenues

Typically, the first component developed in a financial plan is the projection of revenues under existing rates. Revenues are the principal sources of income to the utility. There are three types of revenues common to most utilities: rate or user charge revenue, miscellaneous operating revenue, and nonoperating income such as contributions to capital.

Rate revenue. Rate revenue includes funds received from customers for water service. There are several different methods of forecasting water rate revenue. The preferred method is to determine the revenues by multiplying the current rates by the units of service. Units of service typically include the number of service connections and amount of metered water consumption, including consideration of the expected rate of growth in service connections and metered water consumption for each year of the projection period. Recognizing these factors will allow the utility to more accurately forecast revenue under existing rates, as well as provide consistency with the rate-design process through the use of a common set of assumptions throughout the rate-setting process. This method of forecasting revenues under existing rates is illustrated for the example utility in Table 3-1.

If the data are available, the forecast of rate revenues should be developed by class of service (e.g., residential, commercial) on an annual basis for each year of the forecast period. As indicated in chapter 2, the utility should strive to develop and maintain adequate records of customers by meter size and class of service, as well as consumption data by class, since the projection of these parameters impacts the utility's operations, its income, and ultimately its rates and revenues.

An alternative that may be used to forecast revenue is to apply an appropriate growth rate to historical revenues to project revenue for future years. If the historical revenue method is used, it is important to consider and adjust for variables that impacted past billings and revenues (e.g., weather, conservation, local economy) and, thus, could potentially skew future projections. In addition, any projected changes in rates and the impact on projected revenues need to be taken into account, since the financial plan uses

Table 3-1 Projected rate revenue for the example utility

Line Number		Base	Year 1	Year 2	Year 3	Year 4	Year 5
	Unmetered Accounts*						
1	Residential	200	203	206	209	212	215
2	Nonresidential	40	41	42	43	44	45
3	Total unmetered accounts	240	244	248	252	256	260
4	Current rate per account per month	$25.00	$25.00	$25.00	$25.00	$25.00	$25.00
5	Revenue from unmetered accounts	$72,000	$73,000	$74,000	$76,000	$77,000	$78,000
	Metered Accounts*						
6	Residential	1,900	1,929	1,958	1,987	2,017	2,047
7	Nonresidential	110	112	114	116	118	120
8	Total metered accounts	2,010	2,041	2,072	2,103	2,135	2,167
9	Current rate per account per month	$12.50	$12.50	$12.50	$12.50	$12.50	$12.50
10	Revenue from metered accounts	$301,500	$306,000	$311,000	$315,000	$320,000	$325,000
	Metered Consumption—1,000 gal†						
11	Residential	240,000	243,663	247,326	250,989	254,779	258,568
12	Nonresidential	45,000	45,818	46,636	47,455	48,273	49,091
13	Total metered consumption	285,000	289,481	293,962	298,444	303,052	307,659
14	Current consumption rate per 1,000 gal	$1.50	$1.50	$1.50	$1.50	$1.50	$1.50
15	Consumption revenue	$427,500	$434,000	$441,000	$448,000	$455,000	$461,000
16	Total all-rate revenue	$801,000	$813,000	$826,000	$839,000	$852,000	$864,000

*Based on a 1.5% annual growth rate; varies because of rounding.
†Assumes water use per account in base year continued at the same level in subsequent years.

a projection of revenue at existing or adopted rates as the basis for evaluating the need for rate adjustments.

In the case study example, the base year revenues are estimated at $801,000 as illustrated in Table 3-1. Using 1.5 percent annual growth rates for both the residential and nonresidential accounts and assuming that the base-year water use per account does not change, annual rate revenue is expected to increase by 1.5 percent.

Miscellaneous revenue. This category of revenue includes all miscellaneous fees and charges for service other than water rates (e.g., bad check fees, connect or reconnect fees, meter testing fees, interest earned on invested cash in various funds and accounts). These revenue sources are typically projected based on historical results and any known changes occurring on the system, such as the implementation of a new fee or adjustments to existing fees.

Nonoperating income. This revenue category includes interest generated from invested cash set aside in various restricted funds, such as construction funds, and other revenues collected for specific purposes, such as contributions in aid of construction (contributions) collected from new customers and developers. These contributions may come in the form of system development charges (also known as *impact fees* or *capacity fees*), which are used to fund capacity for new customers. System development charges are one-time charges imposed on new customers of the system and are more fully described in AWWA Manual M1. The utility should refer to its state and local laws as to the appropriate use of these funds. In many jurisdictions, revenues from these charges must be tracked separately from regular rate or miscellaneous revenues because they may only be used to defray capital costs. System development charges can often only be used to pay growth- or expansion-related capital costs or debt service.

Unlike contributions to capital, both rate revenue and miscellaneous revenue are sources of funds that can generally be used to offset any type of utility-related expenditure, whether capital or operating in nature.

Expenditures

A typical starting point for developing a projection of expenditures is a utility's annual operating and capital budgets or audit. Historical costs are also a good source of information and guide in providing a basis for the projections of future cost levels beyond the current budgeted levels. From these historical expenditures, consideration will be given to identifying nonrecurring costs, future anticipated costs such as new employees, or an increase in operation expenditures and/or capital expenditures due to the addition of new or expanded facilities.

O&M expenditures. O&M expenditures include the day-to-day expenditures for such items as labor, materials, utilities (such as power), supplies (such as chemicals), and miscellaneous expenditures. The financial plan typically includes these expenditures on a cash basis, i.e., the financial plan reflects the cash outlay for these items as they are incurred during the fiscal year.

When projecting O&M expenditures, it is important to consider the format that they will be presented in the financial plan. Ideally, these expenditures are maintained in a format that parallels the various water service functions (e.g., source of supply, treatment, pumping, and transmission). Tracking expenditures on a functional basis identifies how much of the total cost is represented by each element of the system, which ultimately simplifies the establishment of cost-of-service–based rates. A standard classification of operation expenditures (called a *chart of accounts*) is published by the American Water Works Association in its *Financial Management for Water Utilities: Principles of Finance, Accounting, and Management Controls.* Also, the National Association of Regulatory Utility Commissioners (NARUC) publishes a similar chart of accounts that is often required by regulating agencies for investor-owned utilities. Regardless of source, these charts of accounts may be used by municipally owned utilities, and if followed, facilitate the functionalization of O&M expenditures into cost categories that are used in the cost-of-service process.

If O&M expenditures are not maintained on a functional basis, the utility could use its existing accounting records and/or budget documents to project O&M expenditures. Small utilities often maintain separate line item expenditures for labor and payroll-related expenses, professional services, power and chemicals, insurance, and miscellaneous expenses. The most important consideration is to separate O&M expenditures into sufficient detail to allow the application of different cost escalation factors or changes in the overall level of certain expenditures as appropriate. For example:

- Labor costs may generally rise in proportion to the overall level of inflation over time (if for no other reason than to cover increases in employee benefit costs) or may be influenced or governed by a collective bargaining agreement.
- Health insurance and pension-related expenditures may increase at a higher rate than labor based on historical trends.

- General or property insurance may increase based on increases to the utility's fixed asset base as a result of the capital program implementation.
- Variable treatment costs such as electricity and chemicals may increase or decrease based on (1) changes in the cost of the commodity and (2) estimated changes in water gallons treated.

It is important to project these expenditures separately from other expenditures that may remain relatively fixed over time or have some other pattern of change that is significantly different from inflationary growth.

Contract payments, for example, are generally set by the terms of the contract and may or may not include an inflationary adjustment. Transfer payments to other city or municipal departments (if the water utility is part of a city or other larger entity) may also have a growth pattern that differs from inflation and as such should be treated separately.

Table 3-2 shows an example projection of O&M expenditures for a 5-year period. In this case, year 1 of the projection period is the current operating budget for the example utility. For projection purposes, an annual inflation rate of 3 percent has been applied to all cost categories with the exception of the maintenance category. In year 3, a $50,000 incremental increase in maintenance is reflected over and above the 3 percent inflation rate. For the example utility, this reflects a change in maintenance activity or program that is permanent (at least for the balance of the study period). Note that the breakdown of O&M expenditures for the example utility is available by expenditure category and not by utility function.

All assumptions used in the development of the multiyear forecast period should be carefully documented so that someone reviewing the projections can check for reasonableness.

In the case study example presented in Table 3-1, the base-year revenues are estimated at $801,000. Using 1.5 percent annual growth rates for both residential and nonresidential

Table 3-2 Summary of projected O&M expenses

Line Number	Category	Year 1	Year 2	Year 3	Year 4	Year 5
1	Payroll and payroll-related	$165,000	$170,000	$175,000	$180,000	$185,000
2	Maintenance	144,000	149,000	203,000	209,000	215,000
3	Professional fees	26,000	27,000	27,000	28,000	29,000
4	Power	31,000	32,000	33,000	34,000	35,000
5	Insurance	21,000	21,000	22,000	23,000	23,000
6	Office and billing	31,000	32,000	33,000	34,000	35,000
7	Education and certifications	5,000	5,000	5,000	6,000	6,000
8	Permits	7,000	7,000	8,000	8,000	8,000
9	Telephone	4,000	4,000	4,000	5,000	5,000
10	Trucks	8,000	8,000	9,000	9,000	9,000
11	Pension plan	10,000	11,000	11,000	11,000	12,000
12	Public notices	2,000	2,000	2,000	2,000	2,000
13	Other utilities	5,000	5,000	5,000	6,000	6,000
14	Total O&M costs	$459,000	$473,000	$537,000	$555,000	$570,000

accounts and assuming that the base-year water use per account does not change, annual rate revenue is expected to increase by 1.5 percent.

Capital expenditures. In general, the capital costs of a utility include the capitalized asset and equipment purchases, renewals and replacements, and that portion of capital construction projects of the utility that are funded directly from annual system revenues on a pay-as-you-go basis or with the accumulation of cash reserves. Such capital costs recovered directly from system revenues, along with the issuance of long-term debt (and potentially the use of grant funds), provide the sources of funding for the utility's capital improvement program (CIP).

In the development of the projection of annual capital expenditures, the utility should use available documents such as master plans, developer-prepared plans, or utility-generated capital improvement programs. If these sources of data are not available, the utility should generate a capital improvement program that defines the future needs of the utility. Ideally, at least a 5-year program should be developed to better anticipate future capital and funding needs.

The CIP should take into consideration a number of factors, including the identification of the years over which various projects are anticipated to be constructed, a general description of each project, and the associated costs. In addition, any new operating costs that are anticipated once the project is complete or online (e.g., personnel to operate a plant or power costs for a pumping station) should be identified. The capital program should also attempt to prioritize the needs of the individual capital projects in any given year. This becomes important if funds or construction time become limited in a particular budget year. For the example utility, a multiyear CIP and funding program is provided in Table 3-3. The objective of this table is to enable the utility to determine the optimal mix of funding sources, displayed under sources of funds, available to complete the capital plan while keeping rate impacts to a minimum.

The funding of the CIP should consider funding sources that may be available to help offset capital costs. Such funding sources may include system development charges, grants, developer contributions, low-interest state revolving loans, and bonds. Bonds can be structured in a variety of ways to meet financing objectives, such as wrapping bond debt service payments around existing debt obligations to make aggregate debt service level. Debt financing can help the utility minimize the initial impact of capital program costs on the utility and its customers. For more information on potential funding sources for the utility's capital program, the reader is referred to AWWA Manual M29, *Water Utility Capital Financing.*

When developing the capital program, it is important to remember that funding the CIP with current revenues causes a direct impact on rate revenues. That is, for every dollar of capital funded from rate revenues, one dollar must be collected from rate revenues. However, once the water rates are established at a level that support a certain amount of CIP funding with current revenues, a similar amount of the CIP can typically be funded each year without the need for additional rate increases for this purpose. In contrast, financing the CIP with long-term debt has a less immediate impact on rate revenues, but the annual rate revenue requirements related to the debt financing typically extend for a long period of time (e.g., 20 years). Grant funding does not typically impact rate revenues, with the exception of when the grant requires matching local funds. In the case study example, the total anticipated annual capital funding required from rates (line 12 in Table 3-3) is carried forward to the example financial plan shown in Table 3-4 line 14, as the rate-financed capital costs component of the financial plan.

Utilities should consider funding capital improvements from current revenues (i.e., on a pay-as-you-go basis) in an amount at least equal to the amount necessary for normal renewals and replacements of utility assets such as pumps, pipelines, valves, and meters. As a practical matter, the utility will need to assess the ability to cash-fund projects

Table 3-3 Summary of 5-year capital improvement program

Line Number		Year 1	Year 2	Year 3	Year 4	Year 5	5-Year Total
	Capital Outlays						
	Water Supply						
1	Well pump, motor, equipment at well*	$0	$0	$150,000	$300,000	$0	$450,000
2	Chemical treatment unit at well site*	0	0	125,000	400,000	0	525,000
	Storage Capacity						
3	0.5-mil gal tank at well site*	0	0	150,000	250,000	0	400,000
4	0.5-mil gal tank for north zone	0	150,000	250,000	0	0	400,000
	Pumping Capacity						
5	Add 500-gpm booster pump at Avenue B site	0	0	0	50,000	150,000	200,000
	Mains						
6	Water main replacements	0	0	0	125,000	125,000	250,000
7	Unidentified	350,000	211,000	0	0	118,000	679,000
8	Total capital outlay	$350,000	$361,000	$675,000	$1,125,000	$393,000	$2,904,000
	Sources of Funds						
9	System development charges	$50,000	$51,000	$52,000	$53,000	$54,000	$260,000
10	Grants	125,000	130,000	346,000	0	0	601,000
11	Loans	0	0	100,000	900,000	162,000	1,162,000
12	Capital funding required from rates	175,000	180,000	177,000	172,000	177,000	881,000
13	Total funding sources	$350,000	$361,000	$675,000	$1,125,000	$393,000	$2,904,000

*This funding source, as illustrated for the example utility (see Table 3-4), is likely to be a loan or line of credit that is accessed on an as-needed basis and does require the utility to achieve a minimum debt service coverage ratio—1.25 times the annual debt service or loan repayment amount.

independent of the amount of annual depreciation expense, because depreciation expense is typically based on the original cost of the fixed assets placed in service in prior years and most likely does not directly relate to the cost of renewals or replacements that are currently needed. Moreover, annual depreciation expense reflects the amortization or loss in value of all of the utility's assets, including major structures and facilities, which may be funded with sources other than direct annual rate revenues.

The phrase *depreciation expense* is often confused with *renewals* and *replacements* in ratemaking. On the one hand, depreciation expense refers to an accounting entry based on the amortization of the original investment (and therefore is often excluded in the revenue requirements for budgeting and ratemaking purposes because it is a noncash expense). Depreciation expense may also not be related to the current physical condition, and typically does not reflect the current replacement cost of the fixed asset(s). On the other hand, renewals and replacements deal with capital expenditures needed to prolong or extend the useful life of the asset(s) or the cost of replacement of the asset(s). Furthermore, depreciation expense is typically excluded from debt service coverage calculations.

The case study example has assumed annual rate revenue funding for renewals and replacements at approximately $180,000 per year. Any capital costs in excess of the amount cash-funded from annual rate revenue will need to be funded through other funding sources, such as system development charges, capital reserves, grants, or debt issues.

Debt service. Debt service is the annual principal and interest payments that the utility pays to service its outstanding debt. By issuing long-term debt (e.g., bonds or loans) a utility spreads the costs of the capital improvement over the repayment period, e.g., typically a 20- to 30-year period, thus matching the cost to both existing and future customers with those customers that will use the facilities or otherwise benefit from the improvement. The annual principal and interest payments (i.e., debt service) must be included in the financial plan. The example in this chapter has assumed the issuance of long-term debt as a tool to smooth out capital costs from year to year.

Frequently, a long-term debt obligation will require the utility to achieve various financial ratios. These ratios and/or other requirements are contained in the debt covenants or bond indenture and should be thoroughly understood by the utility as the financial plan is prepared. One such metric is the debt service coverage ratio. A debt service coverage ratio is a financial requirement to enhance the financial capability of the utility to meet its existing and future debt service obligations. The debt service coverage ratio is typically defined as annual net revenues divided by annual debt service, where net revenues are generally defined as gross revenue minus total O&M expenses (excluding depreciation).[1] Debt service coverage requirements almost always require the ratio to be higher than 1.0, but how much higher is usually defined by the covenants entered into at the time of the bond sale or loan agreement. The minimum requirement for debt service coverage is typically in the range of 1.10 to 1.30 times annual or maximum debt service. Higher coverage requirements are typically associated with revenue bonds as compared to a general obligation debt that is backed or supported by the full faith and credit of the taxing power of the municipality. A revenue bond is generally secured only by the revenues of the water utility. Net revenues after payment of debt service (i.e., the coverage amount) can be used for other system purposes such as funding capital projects (e.g., renewals and replacements).

A utility should consider establishing a target for the debt service coverage ratio above the minimum required (e.g., 1.50 or higher) to provide a cushion or contingency and to maintain a sound financial condition. Publications from credit rating agencies provide guidance on what the rating agencies consider to be best financial management practices with respect to coverage ratios. The financing plan should ideally allow a utility to conform to best financial management practices recognized by the utility industry and credit rating agencies. The plan should also enable the utility to maintain affordable rates and promote intergenerational equity (e.g., future customers should not subsidize existing customers). For more information on developing an appropriate financing plan, reference AWWA Manual M29, *Water Utility Capital Financing.*

Taxes. Tax payments may be an additional expense that should be considered during the financial planning process. These tax payments may include property taxes, payment in lieu of taxes (PILOT), business and occupation tax, state utility tax, or franchise fees that may need to be considered when determining cash flow needs. PILOTs are sometimes used to tax utilities that are otherwise tax-exempt entities. Because PILOTs are essentially taxes imposed on tax-exempt entities, they are illegal in many states. Franchise fees are used by general governments (e.g., cities and counties) to recover costs related to utilities that use rights-of-way, including telecommunications, cable television, electric, natural gas, water, and sewer. Taxes are normally applied to investor-owned utilities and are specific to each jurisdiction. Taxes are often controlled by state or provincial law. As such, an

---

[1] A utility's bond covenants generally define the "net revenue" that may be included in the calcu lation of debt service coverage; the specific definition can vary by utility.

individual utility should determine any applicable taxes and the most accurate means of projecting future tax responsibilities to be included in the financial plan.

Cash Balances

If the financial plan is to be an effective tool in projecting the need for future revenue/rate increases, it should also include projections of cash balances that are available for the water utility's use. Customers and elected officials may question proposed rate increases if the utility appears to be maintaining significant cash balances above established fiscal requirements and targets that could potentially be used to fund capital projects or offset other revenue requirements. For purposes of the case study, the unrestricted beginning cash balance was assumed to be $200,000 (see Table 3-4, line 16). This balance changes over time as the future annual revenues and expenditures change.

The utility should make certain that any cash balances identified in the financial plan as available to fund annual revenue requirements are not restricted to other purposes. It is also important to understand the nature of the cash balances. If extra cash is available in the operating or revenue fund because prior year's water demand exceeded projections and generated surplus sales revenue, then typically those monies should be available for any water utility purpose. However, if extra cash is available because system development charge revenues exceeded projections, those monies may be restricted to specific purposes, such as funding capital projects or associated debt service. Restrictions on the use of system development charge revenues vary by location.

Reserves

Maintaining cash reserves is a common way that utilities mitigate financial risk, plan for future system needs, and satisfy bond covenant requirements. Water utilities that have issued debt to pay for capital assets will often have requirements to maintain reserves that are specifically defined to meet the legal covenants of the debt that was issued. Reserves may be established to fund future capital projects, or as a contingency for unexpected fluctuations in revenues or expenditures, or for the funding of unexpected capital improvement needs.

Reserves may also be established for purposes of rate stabilization. The establishment of a rate stabilization fund may allow the utility to set aside a cash amount that can be used to provide rate stabilization in years when revenues are unusually low or expenditures are unusually high.

Various types of reserves and industry guidelines for setting reserve levels are discussed later in this chapter. For purposes of the example utility, a 25 percent O&M or cash reserve target (three months or 25 percent of annual O&M expenditures) has been established along with a capital reserve equal to approximately one year of planned renewal and replacement infrastructure. The next section of this chapter discusses the establishment of target reserve balances.

ESTABLISHING TARGET RESERVE BALANCES

The level of reserves maintained by a utility is another important financial planning consideration that should be reviewed annually. Many utilities, as well as financial rating agencies and the investment community as a whole, place special emphasis on reserve fund balances that are adequate to serve as a cushion or protection to the financial security for the utility under potentially adverse conditions. Utilities should have a well-defined policy regarding their reserve level goals for two reasons. First, it provides a utility guidance in planning to have adequate funds available to meet its financial obligations in times of varying needs. Second and more important, it provides a framework around which financial decisions can be made to determine when reserve balances are inadequate or excessive and what specific actions need to be taken to remedy the situation.

Utility reserve balances can be thought of as similar to savings accounts. Reserve balances are funds that are set aside for a specific cash flow requirement, financial need, project, task, or legal covenant. These balances may be maintained in order to meet short-term cash flow requirements and minimize the risk associated with meeting financial obligations and continued operational needs under adverse conditions.

It is important to establish financial policies that clearly state the basis for establishing targeted reserve balances. Financial policies should articulate how these balances are established, their use, and how to determine the adequacy of the reserve fund balances. For planning purposes, the reserve balances should be reviewed annually during the budgeting process to monitor current levels and provide for conformance with existing policies and practices. Decisions can be made to maintain, increase, or decrease the reserve balances, as appropriate, depending on the impact of such decisions for the upcoming budget period. A utility should review the basis used to establish reserve balances every 3 to 5 years. This time frame is appropriate given that debt levels and capital infrastructure activity can vary during this time, and changes in this factor have an effect on the appropriate level of reserve balances. This type of review allows for the basis of establishing reserve balances to be changed or modified to better reflect the existing conditions and issues.

GENERAL OVERVIEW OF COMMON INDUSTRY PRACTICES

This section discusses the common approaches or practices used by the water utility industry to establish reserve balances. In this process, it is important for smaller utilities to begin considering the establishment and maintenance of various funds in relationship to their specific needs or areas. The most common reserve balances are usually established around the following four areas: operating, capital, contingency, and debt.

Operating Reserves

Operating reserves are typically established to provide the utility with the ability to withstand cash-flow fluctuations. Another term for this reserve fund balance is a *working capital reserve.* There can be a significant length of time between when a system provides a service and when a customer may pay for that service. A study of the utility's historical cash flow and how its accounts receivable balance has changed over time can accurately quantify the time period between delivery and payment for service. In addition to timing, the volume of cash flow can be affected by weather and seasonal demand patterns. This issue is larger for some water utilities than it is for others.

For those who do not have sufficient information to conduct this type of analysis, various industry standards are in use. A 45- to 90-day (approximately 6 to 13 weeks) O&M reserve is a frequently used industry norm. Many utilities attempt to keep an amount of cash equal to at least 45 days, or one-eighth of their annual cash O&M expenses, in an operating reserve to mitigate potential cash flow problems. However, the reserve level is a policy decision for the utility that is influenced by the specific timing of disbursements and receipt of revenues, the financial comfort level of the utility, and the utility's billing frequency. It may also be dependent on the utility's access to short-term "loans" from other sources. For example, if the utility is part of a municipal organization or district, it may have access to working capital funds that would allow for the maintenance of lower reserves. To the extent these funds are "accessed," the utility should plan for the replacement of such funds.

Capital Reserves

A capital reserve balance is intended to be used to fund capital projects that expand system capacity, comply with regulatory requirements, or replace system assets that have become worn out or obsolete. This reserve commonly consists of resources that are restricted for capital purposes (e.g., system development charges, renewal and replacement funding from rates), and can provide a cushion to absorb capital cost overruns that might otherwise require debt issuance and/or rate increases. A minimum balance for this reserve is often defined based on a percentage (1–2 percent) of the cost of system assets or a rolling-average of planned capital expenditures.

Contingency Fund

The contingency reserve fund balance is essentially "insurance" against unanticipated emergencies or failure of the utility's most vulnerable system components. This reserve provides funding for emergency repairs or replacements, typically those caused by natural disasters such as hurricanes, floods, earthquakes, tornados, fires, and similar natural disasters. One method for determining the balance to maintain in such a reserve is to determine the cost of replacing the most expensive facility of the utility system and reserving an amount equal to that cost, less any potential insurance payments that may be applicable. Another method is to base the level of reserve on the historical records and experience of the utility in dealing with such emergencies and disasters, reflecting the chance of incidence of these situations. Determining the emergency reserve level for a system is also a function of management objectives and overall system reliability. The need for a contingency reserve may be mitigated or even eliminated through the establishment of a close working relationship with local lending institutions and creation of an available line of credit to quickly draw on during such a period.

Debt Reserves

Establishment of a debt reserve is typically legally required by the covenants and indentures of most revenue bonds and may be required by some general obligation bonds. Many state lending programs also have similar requirements. The establishment of a debt reserve provides security to the bondholders that adequate funds will be available to pay the debt service obligations, even under a distressed situation.

The debt service reserve fund is typically established at an amount equal to the lesser of the maximum annual debt service on all outstanding senior parity bonds, or 120 percent of the total annual debt service, or 10 percent of the bond issue amount. Often this amount is included in and funded from the principal amount of the bond issue and serves to secure the debt service payment to bondholders. However, it can also be funded from current-year rate revenues over a period of years, e.g., funded monthly over the first 60 months after the issue date.

Rate Stabilization Fund

Some utilities have established a rate stabilization fund or account of unrestricted cash reserves that can be used to meet revenue requirements while gradually increasing rates over time to the level necessary for the utility to be financially sustainable. A utility may have a financial policy pertaining to the funding of such account.

Many bond indentures establish a "rate stabilization fund" that functions as a debt service coverage management tool. During the years in which a transfer is made to the rate stabilization fund, the utility's revenue for coverage purposes is reduced by the amount of the transfer (i.e., the revenue recognition is deferred). The balances in the rate stabilization fund can be used in future years to improve debt service coverage—any withdrawals from the fund are recognized as revenue for coverage purposes. This type of fund also allows a utility to gradually phase in rate increases and reduce "rate shock."

Establishment of appropriate reserve balances will provide utility management with additional resources to manage the financial position of the utility, ensure financial stability, and mitigate financial risk. The reserves provide a buffer in lean years and give the utility greater flexibility at all times because there is "money in the bank" that can be used to take advantage of opportunities that may arise. The reserve recommendations contained within this discussion should be tailored to the specific needs of the utility. Some of the reserves may not be required or appropriate for all utilities.

SUMMARY OF THE FINANCIAL PLAN

A multiyear financial plan will provide the utility with a view of its anticipated financial position, and will help the utility identify the overall level of revenue adjustment required to cover the future operation and capital needs. The financial plan is used as the framework within which the interrelationship of revenues, reserves, operating costs, and capital costs is analyzed.

Elements of a financial plan include projections of revenues, expenditures (O&M, capital, debt service, and taxes), cash reserves, and key financial ratios, such as debt service coverage. The example financial plan described in this chapter has used the operating and capital budgets for the year 1 projections and reflects a 1.5 percent annual growth rate for water service revenues based on current rates (accounts and consumption) and a 3 percent annual escalator for miscellaneous revenue and O&M expenses (with the exception for the previously discussed maintenance category). These factors were recognized in the projections shown in Tables 3-1 and 3-2, which have been carried forward to the financial plan shown in Table 3-4. The projected financing of the capital improvement program, developed in Table 3-3, has also been carried forward to Table 3-4.

The example financial plan illustrates that revenue increases ranging from 32 percent to 57 percent will be required during the 5-year time horizon. It is important to note that the results of the financial plan are cumulative. That is, any revenue or rate adjustment in a prior year will reduce the amount required in subsequent years. Alternative revenue/rate plans are discussed later in this chapter.

OVERVIEW OF THE REVENUE SUFFICIENCY EVALUATION

A revenue sufficiency evaluation involves the assessment of the adequacy of revenues anticipated to be generated under existing rates in meeting the revenue requirement needs of the utility, and an assessment of the rate revenue adjustments necessary in each year of the forecast period in the case that revenues from existing rates are not adequate. The cash flow projection in Table 3-4 shows the level of annual revenue adjustment required in each year of the 5-year projection period (see Table 3-4, line 22, cumulative revenue and reserve deficiency) for the case study example. The revenue increases shown on Table 3-4 are stated for each year as if no revenue adjustments were made in any year of the projection period.

When assessing the annual rate revenue adjustments needed in each year of the projection period, it is necessary to take into account the effect of revenue adjustments that are proposed to be adopted in prior years of the projection period and their cumulative effect on future year rate revenues. A further discussion of how to assess revenue sufficiency and the need for rate revenue adjustments is provide in the following sections.

Table 3-4 Summary financial plan

Line Number		Year 1	Year 2	Year 3	Year 4	Year 5
	REVENUE					
	Operating Revenues					
1	Service revenue—water sales*	$813,000	$826,000	$839,000	$852,000	$864,000
2	Miscellaneous fees	2,000	2,000	2,000	2,000	2,000
3	Total operating revenues	815,000	828,000	841,000	854,000	866,000
	Nonoperating revenues					
4	Tap/connection fees	10,000	10,000	10,000	10,000	10,000
5	Interest income	10,000	10,000	10,000	10,000	10,000
6	Other revenue	5,000	5,000	5,000	5,000	5,000
7	Total nonoperating revenues	25,000	25,000	25,000	25,000	25,000
8	**Total revenue**	840,000	853,000	866,000	879,000	891,000
	EXPENDITURES					
9	**Total O&M Expenses**	459,000	473,000	537,000	555,000	570,000
10	**Subtotal: Net operating income**	381,000	380,000	329,000	324,000	321,000
	Debt Service					
11	Debt service—existing	150,000	150,000	150,000	150,000	150,000
12	Debt service—new	0	0	9,000	87,000	101,000
13	**Total debt service**	150,000	150,000	159,000	237,000	251,000
14	**Rate financed capital costs**	175,000	180,000	177,000	172,000	177,000
15	**NET INCOME (LOSS) FROM OPERATIONS**	56,000	50,000	(7,000)	(85,000)	(107,000)
16	**Plus: Beginning cash balance**	200,000	256,000	306,000	299,000	214,000
17	**Ending cash balance before reserves**	256,000	306,000	299,000	214,000	107,000
	RESERVES					
18	O&M reserve	115,000	118,000	134,000	139,000	143,000
19	Capital reserve	456,000	456,000	456,000	456,000	456,000
20	**Total reserves**	571,000	574,000	590,000	595,000	599,000
21	**ENDING CASH BALANCE AFTER RESERVES**	($315,000)	($268,000)	($291,000)	($381,000)	($492,000)
	CUMULATIVE REVENUE AND RESERVE DEFICIENCY					
22	(line 21 divided by line 1)	-39%	-32%	-35%	-45%	-57%
	DEBT SERVICE COVERAGE					
23	(line 10 divided by line 13)	2.54	2.53	2.07	1.37	1.28

* Revenues at current rates from Table 3-1; values rounded.

EVALUATING THE SUFFICIENCY OF REVENUES PROJECTED UNDER EXISTING RATES

In identifying the test-year revenue requirements for rate-setting purposes, it is necessary to first evaluate the sufficiency of projected revenues under existing rates. This evaluation is illustrated in the utility example described below. The cash flow projections presented in chapter 2 provide a basis for this evaluation. Line 22 of Table 3-4 shows the percentage adjustments in revenue necessary for each year. Line 23 of Table 3-4 tests the adequacy of revenues under existing rates to provide the required debt service coverage. From these values, it can be seen that the existing revenues, although adequate to achieve the required 1.25 times debt service coverage rates, are not sufficient to establish the targeted reserves.

Line 15 of Table 3-4 shows that if no revenue adjustments are adopted, cash flow surpluses of $56,000 and $50,000 will occur in years 1 and 2, respectively. The debt service on the new debt (e.g., loan or line of credit) used to fund a portion of the CIP begins in year 3 at $9,000 and is $87,000 in year 4 and $101,000 in year 5, as shown on line 12 of Table 3-4. This increase in cash requirements in years 3, 4, and 5 is primarily responsible for cash flow turning negative in these same years, as shown on line 15 of Table 3-4.

Line 22 of Table 3-4 shows that increases in revenue are required due to deficiencies in targeted reserve fund balances, not the overall cash balance, which is positive before consideration of the reserves (see line 17 of Table 3-4). Line 16 shows that the beginning cash balance in year 1 is $200,000 compared to the reserve requirement in year 1 of $571,000, shown on line 20 of Table 3-4. Even when the cash flow surplus of $56,000 in year 1 is added to the beginning balance, the cash balance before reserves on line 17 is $315,000 short of the required reserves, as shown on line 21 of Table 3-4.

In this example, utility management has several choices regarding adoption of rate revenue increases because the example utility is generating a positive cash flow in years 1 and 2. These choices include various strategies regarding rate revenue adjustments that will bring reserves up to the required levels and also correct the cash flow deficiencies that begin to appear in year 3. It is interesting to note that rate revenue increases adopted in the earlier years will provide additional revenue that will mitigate the cash flow deficits in the later years, at least to some extent.

As illustrated above, it is important for utilities to develop a plan of revenue adjustments to provide sufficient revenues to meet both their reserve and cash flow requirements over the projection period. However, an important question facing utility management when establishing a rate plan is: What are the revenue requirements for rate-setting purposes in each year that will provide sufficient revenues within an acceptable level of rate adjustments considering the impact on the utility's customers? This aspect is addressed in the next section.

DEVELOPING A PLAN OF RATE REVENUE ADJUSTMENTS

Utility managers often prefer a plan of revenue adjustments that minimizes the impact on their customers in any one year. This goal requires that the revenue requirements for rate-setting purposes be examined throughout the entire projection period, taking into account the effects of prior-year revenue adjustments on subsequent years' financial performance.

Several approaches can be taken to developing a plan of rate revenue adjustments. First, "just-in-time" revenue adjustments can be determined to provide only the revenue increase (or decrease) required to overcome the revenue deficits (or surpluses) in each year of the projection period. A plan with one large rate revenue adjustment in year 1 and no revenue adjustments thereafter is another option, although often this option is unlikely to be implemented due to its high rate impacts on a single year (e.g., possible "rate shock") and its associated unpopularity among customers and elected officials. Finally, a plan with equal percentage increases in revenue each year may be preferable. This alternative is dependent on the variation in the annual increases required and the necessity of fund balances or lines of credit being available to carry the utility through some years when the levelized annual increases may be insufficient to meet all of the utility's reserve and cash flow requirements.

Other multiyear scenarios of rate revenue adjustments are also possible. The preceding examples were chosen for discussion simply to demonstrate the potential options a utility manager has in order to achieve the objectives regarding a program of rate revenue adjustments while providing the required revenue in each year of the projection period.

In some of the alternative plans discussed, the utility may have to fund a portion of the projected deficits (1) by accepting a lower level of reserves for a period of time, (2) from a line of credit, or (3) from some other source. If such sources of temporary funding are available, those plans may better achieve the objective of reducing rate shock, which may occur when the utility is required to implement a large rate increase in one year.

EVALUATING ALTERNATIVE PLANS FOR RATE REVENUE ADJUSTMENTS

When evaluating alternative plans for rate revenue adjustments, it is important to consider management's objectives regarding annual revenue adjustments. If the cash flow analysis shows that annual revenue adjustments required beyond the initial year are nominal (as in Table 3-4), the actual revenue increase required in the first year of the projection period may be adequate to adjust rates to meet the revenue requirements for the entire period. However, if required revenue adjustments are increasing or are large in any year of the projection period, management may want to explore strategies to normalize the impact of revenue adjustments on the utility's customers over a period of several years. The alternatives that were identified in the previous section are developed as scenarios and applied to the case study in the following sections to illustrate these strategies.

Scenario 1: Just-in-Time Revenue Adjustments

This scenario presents the annual percentage revenue adjustments (increases and decreases) that would be necessary in each year of the projection period if revenues were adjusted each year by just enough to meet the revenue requirements for the year, i.e., produce an ending cash balance after reserves (Table 3-4, line 21) equal to zero. Note that the percentage adjustments discussed in the following scenario have been rounded to the nearest whole percentage for presentation purposes; however, the numerical examples carry the actual percentage adjustments out to more significant figures.

Scenario 1 simply determines the annual revenue adjustments necessary in each year to overcome the percentage deficiency identified on line 22 of the cash flow analysis shown in Table 3-4, taking into account, in each year, the effect of prior year revenue adjustments. This scenario does not include a smoothing of the impacts of revenue adjustments but simply presents the annual adjustments necessary on a just-in-time basis each year.

In the case of the example utility, adoption of a rate increase equal to the deficiency identified for year 1 on line 22 of Table 3-4, or a 39 percent rate revenue adjustment, will provide sufficient revenue to eliminate the deficiency in the cash reserves for the year in question. The increase will also provide more than sufficient revenues in all subsequent years of the projection period to meet all of the projected revenue requirements.

In fact, a just-in-time revenue scenario recognizes both rate revenue increases and decreases, as illustrated in Table 3-5. However, such a scenario is unlikely to be feasible since few governing bodies (i.e., city councils or boards) would knowingly adopt a just-in-time plan of back-to-back rate revenue increases and decreases. The just-in-time scenario provides a starting point to develop other scenarios that could be considered for adoption.

For this reason, two other revenue increase scenarios were developed and are shown in Table 3-5.

It should be noted that debt service coverage, shown on line 23 of Table 3-4, is already adequate in all years of the projection period, under existing rate levels, assuming a debt service coverage requirement of 1.25 for the example utility. For the example utility, revenue increases are not needed to maintain debt service coverage. If debt service coverage was not adequate under existing rate levels with this or any other revenue adjustment plan in any year, higher revenue adjustments would need to be implemented to provide the needed debt service coverage. Therefore, debt service coverage should be monitored in this and all revenue adjustment plans.

This revenue adjustment scenario (just-in-time revenue adjustments) provides sufficient revenue in each year but requires a large increase of 39 percent in one year followed by a 32 percent decrease in year 2. Therefore, if this utility could accept lower levels of reserves in the early years of the projection period, a lower rate adjustment in year 1 could be implemented. This option is evaluated in the following section.

Table 3-5 Summary of alternative revenue increase plans (in thousands of dollars)

Line Number		Year 1	Year 2	Year 3	Year 4	Year 5
	Scenario 1					
	Just-in-Time Revenue Adjustments					
1	Annual revenue adjustment (%)	39%	(32%)	9%	8%	2%
2	Revenue change from cumulative adjustments	$315	($52)	($19)	($86)	($106)
3	Annual cash flow (deficiency)	$371	$3	$16	$5	$4
4	Annual net balance (deficiency)	$0	$0	$0	$0	$0
	Scenario 2					
	Single Increase in Year 1					
5	Annual revenue increase (%)	12%	0%	0%	0%	0%
6	Revenue increase from year-1 increase	$101	$102	$103	$104	$105
7	Annual cash flow (deficiency)	$157	$153	$98	$21	$0
8	Annual net balance (deficiency)	($214)	($64)	$18	$34	$30
	Scenario 3					
	Equal Annual Increases					
9	Annual revenue increase (%)	4%	4%	4%	4%	4%
10	Revenue increase from cumulative adjustments	$30	$62	$96	$131	$168
11	Annual cash flow (deficiency)	$86	$112	$90	$48	$65
12	Annual net balance (deficiency)	($285)	($176)	($102)	($59)	$2

NOTE: Complete alternative cash flow plans are shown in appendix A.

Scenario 2: Single Increase in Year 1

This scenario is an example of accelerating the plan of annual rate revenue adjustments such that a single rate increase in year 1 is sufficient for the entire projection period. As discussed in the just-in-time revenue adjustments scenario for the example utility, this scenario required only a rate increase in year 1 to meet the targeted reserve levels (without a decrease in year 2, this plan would produce cumulative surpluses). However, it is often the case that a utility will not need a rate adjustment until a later year of the projection period. In that case, accelerating the plan of rate adjustments to year 1 and implementing a rate increase in year 1 of sufficient magnitude to provide sufficient revenues in all subsequent years could be considered. Typically, it is unusual to have revenue increases implemented until they are actually needed; however, phasing in, or normalizing, increases or avoiding future rate shocks are valid considerations for implementing an increase earlier than absolutely required.

For the example utility, it is possible to adopt a single rate adjustment in year 1, which is considerably smaller than the just-in-time adjustment discussed in the previous section, provided the utility is willing to accept reserve levels in years 1 and 2 that are somewhat below the targeted levels. In this plan, an overall revenue increase of 12 percent in year 1 will generate sufficient additional revenues over the projection period to result in adequate cash flow in year 5 (line 7, Table 3-5) and reserve balances slightly in excess of required levels in years 3 through 5 (line 8, Table 3-5).

However, utility managers and governing bodies (i.e., city councils, utility boards, and commissions) may prefer to implement several modest revenue increases rather than one large rate revenue increase in a single year that will cover the entire 5-year projection period. With equal annual percentage rate adjustments, rate shock can be mitigated. However, it is important to consider the effects on ending cash balances throughout the period and the ability of the utility to accept lower-than-desired levels of reserves for some years of the projection period. The next section evaluates the option of equal annual percentage rate adjustments.

Scenario 3: Equal Annual Increases

Under this scenario, equal annual percentage revenue increases are applied in each year of the forecast period resulting in the generation of sufficient revenues to meet all revenue requirements by the end of the 5-year projection period. For example, as shown in Table 3-5, equal annual percentage rate adjustments of 4 percent per year are necessary to meet the cumulative projected reserve and cash flow deficiencies under existing rates for the example utility.

Line 10 of Table 3-5 shows that this rate adjustment plan will generate an increasing amount of revenue each year from the cumulative impacts of the annual rate adjustments. Table 3-5 also shows on line 11 that this rate adjustment plan will result in a positive cash flow in each year of the projection period. The only drawback to this rate adjustment plan is that in years 1 through 4, the reserve balances are below targeted reserve levels, as can be seen on line 12 of Table 3-5. However, the deficits in the reserve balances are lower each year and the balance or reserve targets are met by year 5.

This plan results in a higher cumulative percentage increase in rates (21.7 percent)[2] over the projection period than the single increase of 12 percent in year 1. However, it does have the advantage of smoothing the impact of rate revenue adjustments over each year of the projection period. Therefore, if it is the objective of utility management to smooth the impact of rate revenue adjustments over the 5-year projection period, this plan could

[2] The annual rate increases are compounded over time so that the actual increase in rates during the 5-year period exceeds the simple sum of the annual rate increases.

be implemented if (1) the utility could accept reserve balances below desired or targeted levels in the first 4 years of the projection period, or (2) a line of credit or other funding source was available to supplement or fund the reserve fund deficits if needed in the early years of the projection period.

Similar plans could be developed with equal annual percentage increases in revenue for fewer years, say years 1 through 3, with no revenue adjustments thereafter. Such plans would require higher annual revenue adjustments to achieve the same level of annual revenue in fewer years.

Developing Rates to Generate the Revenue Requirements

Once the revenue requirements for rate-setting purposes have been determined for year 1 in the context of the 5-year financial plan projection, there are two approaches to calculating specific rates to generate the required revenue. First, if the current rate structure reflects appropriate cost allocations, is in conformance with generally accepted rate-setting criteria, and little has changed regarding costs, system configuration, and/or customer usage patterns, the current rates may be adjusted appropriately on an across-the-board basis by the percentage adjustment in revenues identified for year 1. For example, all components of the rate—the fixed and volume rates—could be increased 4 percent, as illustrated by Scenario 3. However, if cost allocations or rate structure need revisiting, modifications to the rate design should be considered that are in conformance with generally accepted cost allocation and rate-setting criteria, as discussed in the next chapter, and that will generate the specific revenue requirement for rate-setting purposes for year 1 of the cash flow projection.

Updating the Revenue Requirements Analysis

After establishing a plan of revenue adjustments associated with the 5-year financial plan, it is important to monitor and adjust the plan over time. A good time to review the plan is during the annual budgetary process. The plan can be updated with current information, such as actual beginning-of-period reserve fund balances. The 5-year plan can also be extended—the past year can be dropped; year 1 can include the new proposed budget; an additional year can be added as year 5; and the projections can be updated to include these changes.

If year 1 results were close to the projected results and if the budget for the new year 1 is close to the projections for year 2 in the previous analysis, it may be possible to continue with the same plan of rate revenue adjustments developed in the previous plan. However, if year 1 results and/or the budget for the new year 1 are significantly different than projected, changes may be required to the plan of future annual revenue adjustments.

This page intentionally blank.

Chapter 4

Rate Design

The data and resource constraints faced by a water utility can affect the rate design alternatives available to the utility. This chapter addresses how rates can be designed to generate sufficient revenue to meet the utility's revenue requirements, under the selected revenue requirements alternative developed in chapter 3. This chapter illustrates a rate design that emphasizes simplicity, while at the same time recognizing cost-of-service and equity considerations. However, this simple rate design may not address all of the issues and the unique complications faced by utilities, whether large or small systems. In addition, a simplistic approach to rate design may not meet the utility's rate-pricing goals or objectives. Where potential complications do exist, the reader is referred to the relevant sections of AWWA Manual M1.

The preceding chapters projected revenue that would be generated under existing rates and then developed a pattern, or series of revenue or rate adjustments, that could be used to meet the identified revenue requirements and some or all of the targeted reserves. That analysis is important in that it defines the magnitude of needed revenue, or rate, adjustments over a 5-year or applicable planning period. For the example utility, the scenario selected was Scenario 3, equal annual increases. This scenario is used to illustrate the rate design example included in this chapter.

When reviewing rate design alternatives, a good first step is to determine what is important to the utility. The utility's rate design should send signals to its customers that reflect these priorities. In broad terms a three-step process can be used for selecting a rate design and designing rates:

Step 1: Define rate or pricing goals and objectives.

Step 2: Evaluate rate alternatives.

Step 3: Understand and communicate outcomes.

This chapter begins with an examination of the rate or pricing goals and objectives. It is important for utility management to consider exactly what it wants a rate design to accomplish prior to creating rate design alternatives. Next, basic terminology regarding rate designs is discussed. Fixed and variable charges are separately identified and defined. After this foundation has been laid, an example methodology for the design of rates is identified to meet the example goals and objectives.

Next, rate evaluation issues are reviewed and finally other rate considerations are discussed. Chapter 5 addresses the issue of communicating proposed changes in existing rates to customers and involving the public to achieve better understanding of the process.

RATE OR PRICING OBJECTIVES

The form of the applicable rate design is largely dictated by the goals and objectives of the utility or the community it serves (they may differ). The rate design illustrated in this chapter stresses simplicity and cost recovery. Utilities serving a single customer class and having limited customer and system usage data or other similar characteristics may want to consider the degree to which the rate design can be understood by its customers and the limitations of the billing system. Incorporating other objectives into rate design may require consideration of some of the other more involved rate designs that are presented in AWWA Manual M1.

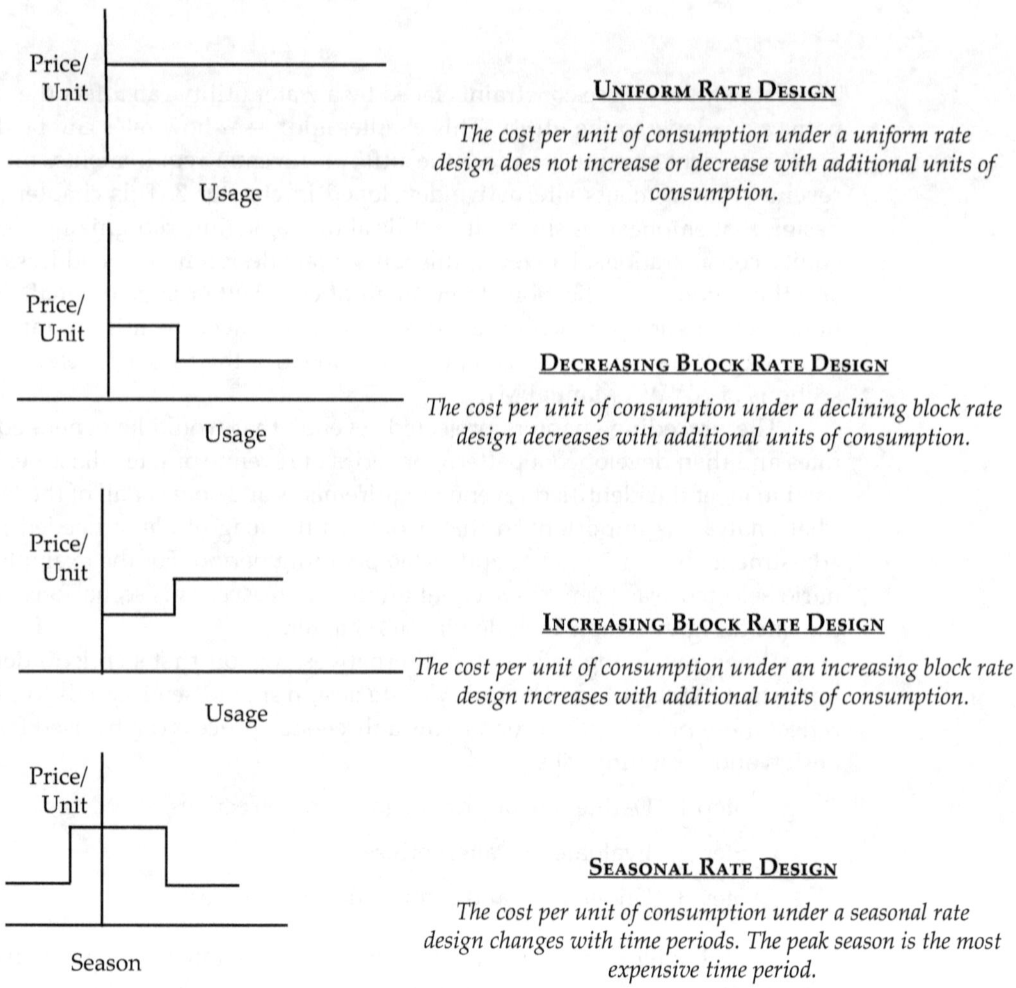

Figure 4-1 Alternative rate forms

Typical objectives that can guide rate design are illustrated in Figure 4-1 and may include the following:

- Revenue sufficiency—the primary objective of any rate design is to recover the revenue requirements or the costs of providing water service.
- Revenue stability—since many costs of providing water service do not vary with changes in water use, utilities often recover a portion of their costs through fixed or base charges per billing period to reduce revenue volatility.
- Simplicity—simple rate designs are often more easily understood by customers; understanding is important to achieving customer acceptance.
- Ease of administration—feasibility and ease of administration are often important concerns. This objective should consider the capabilities of the billing system.
- Affordability—keeping water service affordable to customers enhances the collection of bills as well as revenue stability.
- Resource efficiency—rates sending price signals that encourage efficient water use can enable utilities to avoid or defer costly investments in capacity.
- Legal—the rate designs should be consistent with applicable laws and regulations.
- Consistency with cost-of-service principles—it is generally accepted that those who use the water system should pay for the water system in proportion to the costs they impose on the system. Rate designs consistent with cost-of-service principles are easier to defend and easier for the public to support.
- Fairness—although "fairness" has different meanings to different groups, stakeholders are more likely to support rates that are perceived to be fair.

Although this list is not exhaustive of all the possible goals and objectives, it provides the typical goals and objectives that can be used when evaluating a utility's rate design. In addition, each utility may have a different definition of what the specific goal or objective is. Therefore, each utility should define what its specific goals and objectives are and apply those during the rate design evaluation.

When contemplating rate design goals and objectives, it is important to understand that there are tradeoffs between the various goals and objectives; some of the goals may be mutually exclusive. As a result, it is difficult, if not impossible, to achieve all the goals and objectives in one rate design. Given these tradeoffs, the utility may need to prioritize goals and objectives and design rates that reflect relative priorities. Over time, the utility's goals and objectives may change, which should result in a reevaluation of the rate designs, and if necessary, changes or enhancements to the rate designs to meet the new goals and objectives. Rate design goals and objectives can be achieved through various rate design components. These rate design components are discussed in more detail in the following section.

RATE DESIGN CONCEPTS

A basic distinction of the charges included within a rate design relates to whether the charges are *fixed* or *variable.* This distinction flows from the cost-causation principles—some costs of serving customers do not vary with the amount of water consumed. Nearly all water utility costs are fixed in the short run. Although recovering the vast majority of costs through a fixed charge would certainly make revenues more stable and predictable, this practice may be counter to many other goals and objectives.

A subset of fixed costs is generally referred to as *customer related costs;* these costs typically include meter reading, billing, accounting, collection expenses, and maintenance and capital costs related to meters. These costs are generally recovered through a *fixed* charge per billing period per customer, account, or meter, and do not vary with consumption. Fixed charges may be referred to by many names, such as *customer charges, base charges, meter charges,* or *service charges.* Regardless of the terminology, the fixed charge is in place to recover a portion of the fixed costs the utility incurs to provide service. Costs associated with meter reading, billing, and collection typically are the same for all customers and usually do not vary by water use or the size of the customer's meter. These are often termed a *billing cost component.* Some fixed charges often include costs related to meters and service connections (a *metering component*). These costs do vary and increase as the size of the meter increases.

Because there are many other fixed costs, it is not unusual for some utilities to include other cost elements in the fixed charge. These can include a portion of cost elements such as annual debt service, distribution-related costs, or other fixed system costs. This assessment is typically done to reflect the cost-causation principle that a portion of the system must always be ready to serve customers no matter how demand fluctuates at any given point in time. *Readiness-to-serve* or *demand charges* can be based on the proportional share of providing service for distribution costs and other fixed costs. Readiness-to-serve charges, or any fixed cost that is included in the calculation of fixed charges, will enhance revenue stability but should be carefully considered when reviewing customer impacts.

Another form of a fixed charge is a *minimum charge.* A minimum charge is typically used to refer to a fixed charge imposed until a customer reaches a set amount of consumption to meet the minimum bill requirements. Some utilities may also include a usage allowance within the fixed charge. Regardless of the naming of the charge or the costs to be recovered, the fixed charge is intended to collect the desired level of fixed costs of the utility.

In addition to the above-mentioned fixed charges, utilities may also have a fixed charge for public and private fire protection services. Public fire protection refers to the public hydrants that provide a communitywide benefit. It is not uncommon to find that up to one-third of a small water utility's costs are related to the provision of system capacity for firefighting. Public fire protection costs have occasionally been charged to the local government where the costs are recovered through property taxes in proportion to the value of the property being provided fire protection. However, communities have sometimes been reluctant to pay such charges and, in some areas, public fire protection costs are now being recovered through utility customer rates and charges. Private fire protection refers to the fire protection services provided to specific customers through sprinkler connections and charged to the benefiting customer. Public fire charges are typically assessed through a fee per hydrant, while private fire protection charges are typically assessed based on the size of the sprinkler connection reflecting the capacity or demand requirements should a fire occur. Chapter 4, section 8, of AWWA Manual M1 provides a more detailed discussion of the development of public and private fire charges and rates.

The varying amount of water that customers consume may drive other utility costs. These costs are referred to as *variable costs* and may include power for pumping or treatment, chemicals for treatment, and other costs that may change based on customer consumption. These variable costs (along with the fixed costs not recovered through a fixed charge) are typically recovered through a consumption charge that varies in proportion to the level of a customer's use. The *variable charge* is alternatively referred to as a *commodity charge* or a *volume charge.*

The charge per unit of water consumed is variable in that the component of the bill related to consumption varies based on the amount of water the customer consumes. Similar to the fixed charge, the variable charge can also be collected through a variety of methods. Although this chapter only examines a *uniform rate,* that is, a single charge for each unit of consumption, there are many other forms including block rates and seasonal rates.

Block rates are rate designs where the rate charged per unit of water changes, depending on how much water is consumed. A block of water is a predefined amount of water consumption; for example, the first block might be defined from zero to 15 billing units of water consumption. A block rate may also be referred to as a *tiered rate design* at some utilities. The *declining* or *decreasing* block rate has a volumetric charge that decreases for incrementally larger blocks of consumption. An *inclining* or *increasing* block rate is characterized by a volumetric charge that increases for larger blocks of consumption. *Seasonal* rates are volume charges that change depending on the season of the year. In general, a higher rate is charged during the peak season. Figure 4-1 illustrates these typical variable pricing rate options, and AWWA Manual M1 has a chapter describing each of these rate designs.

Each of the rate designs in Figure 4-1 provides a means to express the utility's goals and objectives to its customers as to the cost of providing service. One of the common goals of rate designs today is providing a conservation incentive. This can be accomplished through several different rate designs. Most commonly a utility will utilize an increasing block rate design, where the rate increases at higher levels of consumption, to discourage wasteful uses. Key to the development of the increasing block rate design is how the blocks and pricing of the blocks are established. The establishment of the block or tier sizes should reflect the specific utility's conservation goals and customers' consumption characteristics. Many utilities will base the block sizes on typical, or average, consumption for indoor and outdoor use, and adjust the size to reflect the goals established for water consumption. A seasonal rate design can also provide a conservation signal to customers and still maintain a more simplistic approach. Regardless of the utility's goals and objectives, the establishment of appropriate rate designs will help meet those goals and objectives. For a more detailed discussion of the various types of water rates designs, the strengths and weaknesses of each design, and how they can be established, please refer to the rate design chapter in AWWA Manual M1.

Rate Design Example

Chapter 3 discusses the determination of projected revenue requirements that extend over a 5-year financial planning period. The projection of the revenues under the existing rate design indicates that projected revenue is less than the revenue requirements and that a revenue, or rate, adjustment is necessary. The following section of this chapter discusses the development of projected rates for the 5-year study period, using the equal annual increases scenario identified in Table 3-5, with a 4 percent increase in rate revenue for each year. The proposed approach to rate design shown in this chapter is to separately examine and establish fixed and variable components of the rate design.

Fixed charges. To start the development of the proposed rate design, it is important to determine if the costs have changed materially for the fixed or customer-service-charge portion of the existing rate schedule. To answer this question, the utility needs to examine those costs associated with fixed charges. While these costs will vary from utility to utility, in general terms, including only meter reading, billing, accounting, collection expenses, and maintenance and capital costs related to meters would yield an estimated fixed charge that might range from as low as $5.00 to as much as $15.00 per account per month. Fixed charges may include more than those specific items listed above. As an example, fixed charges might also include recovery of costs incurred to provide public fire protection, readiness to serve, and so on. As previously noted, it is not unusual for small utilities to have higher fixed charges as a means of enhancing revenue stability or for other reasons relating to customer acceptance and understanding.

For the sake of simplicity in this example, it is assumed that there have been no substantive changes in the costs included in the fixed charge, other than inflationary increases. Accordingly, the existing fixed charge of $12.50 per month for metered accounts has been

Table 4-1 Illustration of projected revenue from fixed service charge

Line Number		Base	Year 1	Year 3	Year 3	Year 4	Year 5
1	Metered accounts*	2,010	2,041	2,072	2,103	2,135	2,167
2	Number of bills per year	12	12	12	12	12	12
3	Annual bills	24,120	24,492	24,864	25,236	25,600	26,004
4	Meter charge†	$12.50	$12.88	$13.27	$13.67	$14.08	$14.50
5	Revenue from metered accounts	$302,000	$315,000	$330,000	$345,000	$361,000	$377,000
6	Unmetered accounts*	240	244	248	252	256	260
7	Number of bills per year	12	12	12	12	12	12
8	Annual bills	2,880	2,928	2,976	3,024	3,072	3,120
9	Unmetered account charge†	$25.00	$25.75	26.52$	$27.32	$28.14	$28.98
10	Revenue from unmetered accounts	$72,000	$75,000	$79,000	$83,000	$86,000	$90,000
11	Total fixed service charge revenue	$374,000	$390,000	$409,000	$428,000	$447,000	$467,000

*Future growth projected at 1.5% per year.
†Future increases in charges projected at 3.0% per year.

increased only for inflation over the 5-year analysis period. A similar assumption is made with regard to the current $25.00 charge to unmetered accounts (see Table 4-1).

Another simplifying assumption for purposes of this example is the use of a fixed charge that is the same for all accounts regardless of meter size or customer class. As discussed earlier, the costs of meter reading, billing, accounting, and collection expenses do not vary by meter size; if these were the only costs included in the fixed charge, they would result in a uniform charge for all accounts. However, meter maintenance expenses, the capital costs associated with meters, and other fixed cost components do vary by meter size and their inclusion would result in a fixed charge that increases for larger meters. If other system costs were included in the fixed charge, these might be "allocated" to the various meter sizes using equivalent meter capacity or flow ratios and result in even greater differences in the fixed charge between the meter sizes. In systems that have relatively few customers served with the similar meter sizes, a single fixed charge is sometimes used and is reflected in the rate calculations for the example utility.

Table 4-1 projects revenues to be generated over the 5-year period from the fixed service charge as applied to both metered and unmetered accounts.

Variable charges. Once the fixed charge rates and revenues have been calculated, the next step is to determine the variable charge. To develop the variable charge, the fixed charge revenues are deducted from the total revenue requirements after considering non-operating revenues. This determines the amount to be generated from the variable or volumetric rate (see Table 4-2). The total revenue requirements on line 1 of Table 4-2 is the amount of revenue needed to achieve the financial performance results under Scenario 3, equal annual increases. Table 4-2 begins with this value and deducts other revenues (line 2) and the revenue to be generated from the fixed service charge. The resulting balance, line 4 of Table 4-2, is the revenue to be generated from the volume rate. Dividing this value by the projected volumetric sales projection, line 5, produces the volume rate per 1,000 gal of metered water use—a rate of $1.50 per 1,000 gal (the current rate) in the base year, increasing to $1.85 per 1,000 gal in year 5.

Table 4-2 Illustration of volumetric/usage charge calculation

Line Number		Base	Year 1	Year 3	Year 3	Year 4	Year 5
1	Total revenue requirements*	828,000	$870,000	$915,000	$963,000	$1,012,000	$1,063,000
2	Less: Nonoperating revenue†	(27,000)	(27,000)	(27,000)	(27,000)	(27,000)	(27,000)
3	Less: Fixed service charge revenue‡	(374,000)	(390,000)	(409,000)	(428,000)	(447,000)	(467,000)
4	Net revenue requirement from volume sales	427,000	$453,000	$479,000	$508,000	$538,000	$569,000
5	Volumetric sales in 1,000 gal§	285,000	289,481	293,962	298,444	303,052	307,659
6	Volumetric rate per 1,000 gal	$1.50	$1.56	$1.63	$1.70	$1.78	$1.85
7	Annual percentage change in volume rate		4.3%	4.2%	4.5%	4.3%	4.2%

*Total revenue requirement amounts to revenues under existing rates plus nonoperating revenue (see appendix A, Table A-3).

†From Table A-3, lines 2 and 7.

‡From Table 4-1, line 11.

§Future growth projected at 1.5% per year.

RATE EVALUATION

Rate evaluation is the process of measuring a particular rate design against the desired objectives to see how it complies with these objectives. An example of this process is illustrated in Figure 4-2. Because elements of the objectives and the "ranking" are subjective, different audiences will likely arrive at different weightings or results as compared to those illustrated in Figure 4-2.

As noted above, this is an example of how the evaluation could work. Each utility will have its own definition of each of the objectives, as well as its own ranking of each objective. Many of the goals and objectives can be achieved regardless of the rate design. These are items such as revenue sufficiency (collecting the necessary revenue to operate the utility), legal (meeting all local legal requirements), and cost-of-service based (reflect the costs of providing service). How well the rate design meets the other goals and objectives will provide the utility with the information necessary to implement a rate design to best meet its goals and objectives.

Once a rate design has been evaluated, the utility should develop a bill impact analysis. A bill impact analysis is defined as the difference in a customer's bill under the existing design versus the proposed rate design at various levels of consumption. The review of the bill impact may result in a shifting of the goals and objectives if the proposed rate design results in significant impacts to customers. If significant rate changes are anticipated, public involvement or a rate phase-in approach, discussed in the following chapter, will also take on heightened importance.

Since stability of revenue relative to costs is usually a major concern, it is wise to perform at least a cursory sensitivity analysis by substituting high and low estimates for anticipated sales to examine the resulting revenue impacts. Second, an adaptive approach is to consider more frequent rate adjustments, that is, annual smaller increases instead of less-frequent larger increases. Routine rate adjustments give the opportunity for midcourse corrections due to changing economic and financial conditions.

Objectives	Uniform Rate	Decreasing Block Rate	Increasing Block Rate	Seasonal Rate	Flat Rate
Revenue sufficiency	↔	↔	↔	↔	↔
Revenue stability	↑	↑	↓	↓	↑
Simplicity	↑	↓	↓	↔	↑
Ease of administration	↑	↓	↓	↔	↑
Affordability	↔	↔	↔	↔	↓
Resource efficiency	↔	↓	↑	↑	↓
Legal	↔	↔	↔	↔	↔
Cost-of-service based	↔	↔	↔	↔	↔
Fairness	↔	↔	↔	↔	↔

Legend: ↑ = Increase ↔ = Neutral ↓= Decrease

Figure 4-2 Example of evaluation

OTHER CONSIDERATIONS

There are many other items a utility will want to consider when setting rates. These can include rate goals and objectives as well as possible other charges and fees to recover specific costs. Provided below is a brief discussion of other considerations.

Rate Affordability

A key concern for utilities when increasing rates is the impact on customers' bills (the affordability of the rates). As rates increase over time, affordability of utility rates is becoming a more sensitive issue. However, defining affordability is not always straightforward.

One of the most common measures of rate affordability is as a percentage of median household income (MHI). Based on criteria in publications from credit rating agencies and government agencies such as the US Environmental Protection Agency and state departments of environmental protection, the range of affordability is typically between 1.5 percent and 2.5 percent of MHI per utility bill (each for both water and wastewater). In other words, each utility bill should not make up more than 1.5 percent–2.5 percent of the MHI of the customer base. However, one shortfall of using MHI is that by definition half of the customers will have a household income below the median and as a result will have a higher percentage of their household income used to pay their utility bill. Another issue is that the MHI does not take into consideration that customers with the lowest household incomes, especially those on fixed or declining incomes, are impacted the most.

Other measures of customer affordability have also been developed by specific utilities for their customers. These other measures can include a demographic review of the utility's customer base and determining the impact on various ranges of customer MHI or discretionary income. Regardless of the method to review rate affordability, addressing affordability is one of the key criteria when evaluating the level of rates or the impact of a new rate design to customers.

One method to address a portion of the affordability issues is through the design of the rates. This can be accomplished through both the fixed and variable charges. As mentioned previously, the fixed charge should be set at a level that reflects a portion of the fixed costs. However, simply reducing the fixed charge may not be the answer as this may have a negative impact on the revenue stability and sufficiency of rates. A variable charge that provides a "lifeline" rate, or an initial block of water to provide essential needs at a lower cost, is one method that has been used to address affordability concerns. All customers are provided a lower cost of water to meet essential water needs.

Other utilities may also implement a low-income rate or discount on the bill if a customer meets income-based qualifications, while some utilities establish other funding mechanisms either through customer contributions or funds from other programs. When implementing a low-income rate or other funding mechanisms, the utility should consult its legal counsel to ensure compliance with local legal requirements. Theoretically, low-income rates or discounts should not be universally applied to all customers; they should only be offered to customers that can demonstrate a need. However, this may prove administratively burdensome. For a more detailed discussion on affordability refer to AWWA Manual M1.

Rates by Customer Class

In presenting the principles and guidelines for determining rates for water service for small systems in this manual, it has been assumed that a simple set of rates will be developed that will apply to all customers. This is common practice for small systems because such systems are often faced with resource and data limitations and may have a customer base that is predominately a single class, e.g., residential. In such cases, it is often not practical to develop a more complex rate design with rates by customer class unless there are specific circumstances that necessitate doing so. For example, where a water system's customer base consists entirely of residential customers and small commercial customers, developing rates by customer class is not likely to be either necessary or cost justified. However, if a system has a customer or customers with unique water usage characteristics or if it serves customers outside its municipal boundaries, developing rates to recognize the different characteristics of those customers may be appropriate. When a system is faced with the need to develop more complex rate designs with rates for different classes of customers, AWWA Manual M1 should be consulted.

System Development Charges

System development charges (SDCs), also referred to as *connection fees, tap fees, capital recovery fees, impact fees, plant investment fees,* or *capacity fees,* are charges to new customers connecting to the system to recover a proportionate share of the capital costs to construct the capacity to provide service. Providing service for future growth often requires substantial upfront investment in infrastructure, typically resulting in existing customers funding the costs of infrastructure prior to future customers connecting to the system. This is due to the timing of when the facilities need to be constructed relative to when the growth that will utilize the new capacity will actually occur. This cash flow problem can be very significant for a small system. This issue of the responsibility for growth is one of benefit and risk: Who benefits when and who is at risk if the growth does not occur? In line with cost-causation principles, development is commonly required to pay a proportionate share of the upfront cost associated with growth. The imposition of SDCs can help to support a "growth paying its own way" policy.

Requiring upfront payments means that general rules, guidelines, and policies should be established by the utility so that the development community understands that it will be required to financially bear some of the costs and risks of development and is aware of the specific cost obligations. This allows members of the development community to make the proper business decisions regarding the economics of their business plans and ensures consistency between development cases. A utility should also establish policies related to how the growth-related revenues will be accounted for, typically in a separate fund, and what the funds may be used for (e.g., only for growth-related projects or debt service related to previously financed growth projects). When establishing these charges, a legal review is necessary to ensure that the local laws allow for these charges and that the fees are properly implemented. Many states have statutes and case law governing the imposition, collection, and use of SDCs.

There are several generally accepted methods available to utilities to establish growth-related charges. Most of the time, and as discussed in chapter 2, SDCs are charged on an equivalent residential unit (ERU) basis. This growth-related fee should reflect the reasonable level of infrastructure capital costs per ERU associated with providing service. The *rational nexus* provisions of case law dictate that the charges should represent reasonable estimates of the impact, or demands, that the new customer places on the system facilities. The amount of the SDC should link to the approximate cost of the benefit received by the customer paying the fee. It is important for growth-related fees to be updated frequently to reflect the changes in system infrastructure costs and needs to provide service to new customers. A more detailed discussion of SDCs, as well as an example of how to calculate and implement such fees, is provided in AWWA Manual M1.

Other Miscellaneous Charges

Other miscellaneous charges are developed by utilities to equitably assign, or charge, customers that are benefitting from a specific service provided by the utility. At the same time, this shelters ratepayers from subsidizing the cost of these special services. To maintain equity, customers who utilize miscellaneous services should pay the costs incurred by the utility in providing those services. These costs may include labor and benefits, overhead costs, equipment rental, taxes, and so on. By charging for a particular service, the utility can also better manage the demand for that service. For example, this can best be illustrated when, in the absence of any fees or charges, there may be an overuse of a service (e.g., turn-on/turn-offs). By managing demand for a special service, the utility can more effectively manage its operating costs. Ultimately, service charges for special services are a reasonable way to minimize costs of the utility for the majority of customers, while allocating costs to customers who are directly benefitting.

There are various different miscellaneous charges in place at utilities today. Some examples of the more common charges are provided below.

- Turn-on/turn-off fees
- Inspection fees
- Account setup fees
- Nonsufficient funds fees
- Installation fees
- Plan review fees
- Late payment fees
- Damage fees
- Deposit fees
- Meter testing fees

A utility choosing to implement other miscellaneous charges should work with its legal counsel to determine the legality of any charge prior to implementation. In addition, the utility should ensure that the service is not being charged to the customer through some other mechanism or department. To best reflect the cost of providing the service, these fees should be reviewed and updated from time to time to reflect changing utility costs and inflation.

Chapter 5

The Rate Adoption Process

STRATEGIC AND PROCEDURAL CONSIDERATIONS

Before initiating the rate adoption process, it is important for a utility to plan strategically and to consider procedural matters, notification requirements, and so on. More recently, the awareness of and opposition to utility rate increases have been heightened. As a result of this heightened awareness, it is even more important to develop and implement a thoughtful plan of public/stakeholder outreach and participation. This planning is true for public/local government as well as investor-owned utilities. Although the rate filing process for a regulated utility (whether publicly owned or investor owned) is likely to be far more formal and structured than an unregulated utility, it is nonetheless important to plan for and anticipate the need to actively engage the public in order to increase the likelihood of a successful rate filing.

One of the key components of planning a rate adjustment is communication. In advance of any formal communication or filing of a rate adjustment or increase, a utility should internally build a case. This process, often led by the utility's public information officer (PIO) or someone in an equivalent position, embodies the strategies and specific tactics relevant to the utility's customer base and service area—strategies and tactics that can lead to an open and effective rate adoption process.

A utility's need for a rate increase should be communicated, at the earliest opportunity, to the public, elected officials, and even regulators. By discussing the need for a revenue increase informally, possible impediments to the increase can be avoided. However, utilities should ensure beforehand that such informal communications do not violate any state and/or local laws or rate filing requirements (especially in the case of a regulated utility).

Prior to official announcement of the need to increase rates, utilities should ensure that all rate filing requirements have been met as well as relevant state and local laws/regulations and, for regulated utilities, all rate filing procedures. The utility's need for a revenue increase should be well documented and articulated by those speaking on behalf of the utility. And as previously noted, the utility should strategically consider the level of public involvement that will be necessary. Public involvement is discussed in greater detail later in this chapter.

RATE CHANGE PHASE-IN PLANS

It is important to remember that a proposed rate adjustment has a limited life expectancy—no rate schedule is proposed to last indefinitely. As the costs and/or demands change under which the proposed rates were based, the rates must also change. Furthermore, if the proposed rates are a significant increase from the existing rates, it might be prudent to implement the change in rates over a phase-in period.[1]

For instance, the first increase in rates may be set so that there is a relatively smaller effect on the typical customer's bill. Each subsequent phase-in would increase the rates until the targeted level of revenue is achieved. In the case of implementing new system development charges, a phase-in to achieve full cost recovery over a period of years may be considered. Even if the final level of the charge is known, some policymakers may choose to phase in the increase over several years because of the business planning cycle. This may allow the development community to have better knowledge of the date and amount of increases in advance so that development can proceed in an orderly fashion. It is important that policymakers recognize the cost to current ratepayers associated with reducing system development charges.

An alternative view regarding the phase-in of rate structure changes is worth noting. If the structure is being modified so as to achieve greater water use efficiencies, i.e., conservation, a phase-in plan may actually nullify or mitigate the desired outcome—actually reduce water use. A rate structure that is, by design, phased-in is intended to allow customers to "adjust" or become "accustomed to" the structure change and thus not achieve the stated objective of reduced water use.

REGULATORY APPROVAL OF RATES

Responsibility and authority for the regulation of water utility rates differ depending on the ownership of the utility. In the case of investor-owned (private) water utilities, rates are normally regulated by the state or provincial utility commission. (Depending on jurisdiction, this commission may be called the *public service commission, public utility commission, board of public utilities, or corporation commission.*) Government or municipally owned (public) water utilities may also be regulated by the state or provincial utility commission under certain circumstances, such as when they serve customers outside their corporate boundaries. However, in most instances, the rates of public water utilities, especially small systems, are subject to the review and approval of a local authority such as the municipal or city council, board, water district, or commission.

Regardless of the specific agency or forum responsible for approving a particular utility's water rates, it is important to be familiar with the specific requirements for the approval of those rates. Procedural requirements such as public notice, public input via

---

[1] One recommended source that can assist the reader in identifying and addressing rate shock and affordability issues is the American Water Works Association publication Avoiding Rate Shock: Making the Case for Water Rates.

hearings or written comments, council votes, or referendums must be followed to ensure that rates are not subject to legal challenge on procedural grounds. If the requirements for a given utility are complex or it is certain that the utility is subject to regulation by a state or provincial utility commission, it is advisable to obtain assistance from legal counsel in preparing for the proceedings related to a rate adjustment.

CUSTOMER AND STAKEHOLDER INVOLVEMENT

Early involvement with customers, stakeholders, elected officials, and appropriate agency staff in the rate-making process not only is desirable but also is often critical to successful rate implementation. Early involvement will provide an opportunity to generate understanding and help foster public acceptance of the water rate proposal. It is common for community laws or regulations to require public notice of proposed rate changes as well as a public hearing. For the customer and stakeholders, the worst time to first hear about a proposed water rate adjustment is during a public hearing. Likewise, utility officials and staff would prefer to hear about potential rate structure problems and challenges prior to the hearing. In many cases, such problems and challenges will be those of perception as opposed to those of fact. Early involvement comprising consistent messaging, community meetings, and workshops is typically productive and less confrontational compared to the often highly charged atmosphere of a public hearing.

Commonly public involvement includes elected official workshops as well as meetings or workshops with a cross section of customers and other community stakeholders. For the meetings to be effective, an effort should be made to anticipate how the proposed water rate changes will affect each of the stakeholder's constituencies. Typically, they will want to know how the proposed rate adjustment will affect their particular account, homeowner's association, apartment complex, or business. When interacting with these groups, it is often appropriate and meaningful to begin with a careful explanation of the reasons for the rate adjustment and end the discussion with a comparative analysis of the proposed rate increases. The analysis could include a number of point-of-reference variables such as inflation, current and proposed water bills within customer classes, and proposed

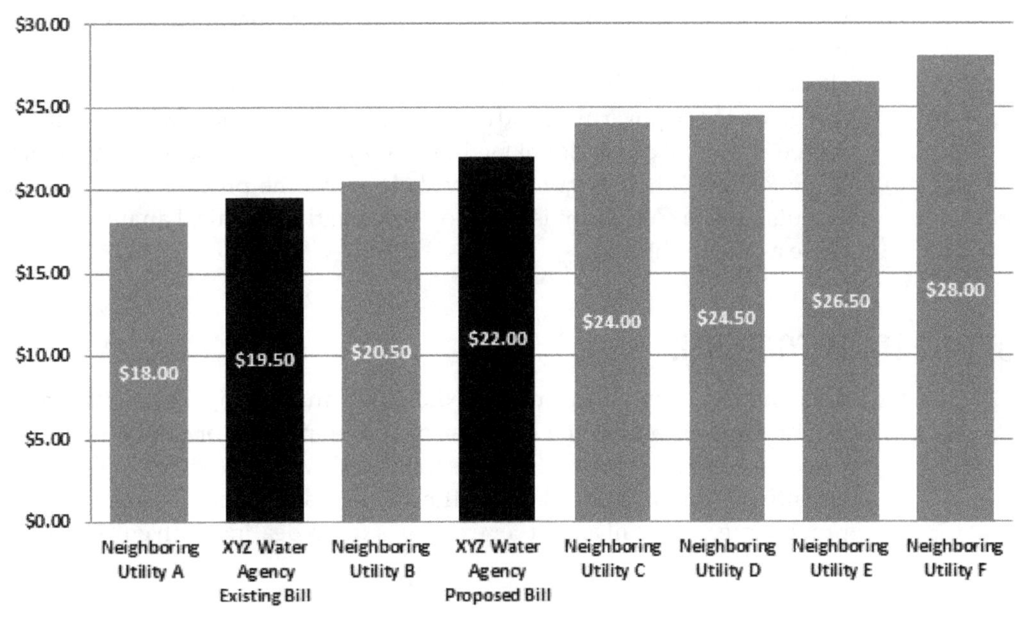

Figure 5-1 Comparison of monthly residential water bills

water bills compared to like or neighboring utilities. An example of an interagency bill comparison is illustrated in Figure 5-1.

INTEREST GROUP MEETINGS

In meetings with different interest groups, it is important to try to determine what their concerns will be ahead of time. This knowledge will provide two benefits. First, presentations can be tailored to address their concerns and issues. Second and more important, the knowledge and the empathy displayed will demonstrate that the utility has made an effort to understand and address their concerns.

Another commonly used approach to public involvement is one in which the utility's policy-making body appoints several representatives to a citizens' or rate-payer advisory committee. The membership of such a committee should be large enough to ensure that all major potential stakeholder segments are represented. Management of this type of committee is very time-intensive; such a committee should typically only be used when significant changes to a utility's rates and charges are being considered. The committee process generally has three phases: education, review, and decision.

During the education phase, time is spent educating the committee on the issues being faced and creating the need for a rate change. It is important that the committee has a clear understanding of the current situation so it can effectively evaluate the alternatives that are being presented.

The review phase is one in which the committee starts to review the different proposals to evaluate the impact on various customer groups using the background information committee members have been given. This involves questions related to the "who, how much, and when" analyses.

Finally, the decision phase occurs when the committee has been provided the necessary background information, has had the opportunity to analyze the proposals, and is prepared to make a recommendation. It is always desirable to have a consensus recommendation, but sometimes this is not possible. If consensus is lacking, it is very important that the different points of view are represented in a clear way to the governing body that will make the final decision. It might be helpful to have an advocate for each position present his or her views. With a citizens' advisory committee, it is also important to establish early in the process the level of input the committee will have into the rate-making changes. Typically, the committee's role is simply advisory, with the final decisions being delegated to the governing body of the utility. In instances when a citizens' advisory committee is used, a professional facilitator may help the process. The facilitator would be an "independent" third party who can help move the process to an effective conclusion. If an outside rate consultant is used to complete the technical analysis, this person may also serve as the facilitator.

PUTTING IT ALL TOGETHER

This chapter is by no means an exhaustive analysis or list of available strategies that a utility could employ when considering a rate increase or rate structure change. Table 5-1 presents a listing of some strategies that a utility may want to consider when defining and executing a public input plan. Furthermore, many resources are available to utilities that provide greater detail and other examples when dealing with customers and stakeholders.

Table 5-1 Basic strategies to winning public support

- Plan ahead and prepare to be responsive.
- Develop mechanisms to receive feedback from customers and other stakeholders.
- Develop a process that does not "hurry to create a failure" but is "patient to create a success."
- Develop an outreach program and identify ways to implement it.
- Let the public help develop solutions through a formal process.
- Ensure elected officials and staff are "on the same page" with proposal points.
- Compare customer impacts within your system and with neighboring agencies.
- Continue to market utility system successes after rates have been adopted.

Appendix A

Alternative Cash Flow Plans

Table A-1 Alternative cash flow plan—just-in-time revenue adjustments

Line Number		Year 1	Year 2	Year 3	Year 4	Year 5
	REVENUE					
	Operating Revenues					
1	Service Revenue—Water Sales	$813,000	$1,146,000	$791,000	$875,000	$955,000
1a	Proposed Rate Adjustment	38.73%	-32.05%	9.00%	7.67%	2.11%
1b	Additional Revenue From Rate Adjustment	$315,000	($367,000)	$71,000	$67,000	$20,000
1c	Service Revenue—Water Sales	1,128,000	779,000	862,000	942,000	975,000
2	Miscellaneous Fees	2,000	2,000	2,000	2,000	2,000
3	Total Operating Revenues	1,130,000	781,000	864,000	944,000	977,000
	Nonoperating Revenues					
4	Tap/Connection Fees	10,000	10,000	10,000	10,000	10,000
5	Interest Income	10,000	10,000	10,000	10,000	10,000
6	Other Revenue	5,000	5,000	5,000	5,000	5,000
7	Total Nonoperating Revenues	25,000	25,000	25,000	25,000	25,000
8	**Total Revenues**	1,155,000	806,000	889,000	969,000	1,002,000
9	**Total O&M Expenses**	459,000	473,000	537,000	555,000	570,000
10	**Subtotal: Net Operating Income**	696,000	333,000	352,000	414,000	432,000
	Debt Service					
11	Debt Service—Existing	150,000	150,000	150,000	150,000	150,000
12	Debt Service—New	0	0	9,000	87,000	101,000
13	**Total Debt Service**	150,000	150,000	159,000	237,000	251,000
14	**Rate Financed Capital Costs**	175,000	180,000	177,000	172,000	177,000
15	**NET INCOME (LOSS) FROM OPERATIONS**	371,000	3,000	16,000	5,000	4,000
16	**Plus: Beginning Cash Balance**	200,000	571,000	574,000	590,000	595,000
17	**Ending Cash Balance Before Reserves**	571,000	574,000	590,000	595,000	599,000
	RESERVES					
18	O&M Reserve	115,000	118,000	134,000	139,000	143,000
19	Capital Reserve	456,000	456,000	456,000	456,000	456,000
20	**Total Reserves**	571,000	574,000	590,000	595,000	599,000
21	**ENDING CASH BALANCE AFTER RESERVES**	$0	$0	$0	$0	$0
22	**CUMULATIVE REVENUE AND RESERVE DEFICIENCY (line 21 divided by line 9)**	0%	0%	0%	0%	0%
23	**DEBT SERVICE COVERAGE (line 10 divided by line 13)**	4.64	2.22	2.21	1.75	1.72

Table A-2 Alternative cash flow plan—single increase in year 1

Line Number		Year 1	Year 2	Year 3	Year 4	Year 5
	REVENUE					
	Operating Revenues					
1	Service Revenue—Water Sales	$813,000	$929,000	$944,000	$958,000	$971,000
1a	Proposed Rate Adjustment	12.40%	0.00%	0.00%	0.00%	0.00%
1b	Additional Revenue From Rate Adjustment	$101,000	$0	$0	$0	$0
1c	Service Revenue—Water Sales	914,000	929,000	944,000	958,000	971,000
2	Miscellaneous Fees	2,000	2,000	2,000	2,000	2,000
3	Total Operating Revenues	916,000	931,000	946,000	960,000	973,000
	Nonoperating Revenues					
4	Tap/Connection Fees	10,000	10,000	10,000	10,000	10,000
5	Interest Income	10,000	10,000	10,000	10,000	10,000
6	Other Revenue	5,000	5,000	5,000	5,000	5,000
7	Total Nonoperating Revenues	25,000	25,000	25,000	25,000	25,000
8	**Total Revenues**	941,000	956,000	971,000	985,000	998,000
9	**Total O&M Expenses**	459,000	473,000	537,000	555,000	570,000
10	**Subtotal: Net Operating Income**	482,000	483,000	434,000	430,000	428,000
	Debt Service					
11	Debt Service—Existing	150,000	150,000	150,000	150,000	150,000
12	Debt Service—New	0	0	9,000	87,000	101,000
13	**Total Debt Service**	150,000	150,000	159,000	237,000	251,000
14	**Rate Financed Capital Costs**	175,000	180,000	177,000	172,000	177,000
15	**NET INCOME (LOSS) FROM OPERATIONS**	157,000	153,000	98,000	21,000	0
16	**Plus: Beginning Cash Balance**	200,000	357,000	510,000	608,000	629,000
17	**Ending Cash Balance Before Reserves**	357,000	510,000	608,000	629,000	629,000
	RESERVES					
18	O&M Reserve	115,000	118,000	134,000	139,000	143,000
19	Capital Reserve	456,000	456,000	456,000	456,000	456,000
20	**Total Reserves**	571,000	574,000	590,000	595,000	599,000
21	**ENDING CASH BALANCE AFTER RESERVES**	($214,000)	($64,000)	$18,000	$34,000	$30,000
22	**CUMULATIVE REVENUE AND RESERVE DEFICIENCY (line 21 divided by line 9)**	-23%	7%	0%	0%	0%
23	**DEBT SERVICE COVERAGE (line 10 divided by line 13)**	3.21	3.22	2.73	1.81	1.71

Table A-3 Alternative cash flow plan—equal annual increases

Line Number		Year 1	Year 2	Year 3	Year 4	Year 5
	REVENUE					
	Operating Revenues					
1	Service Revenue—Water Sales	$813,000	$856,000	$902,000	$950,000	$999,000
1a	Proposed Rate Adjustment	3.73%	3.73%	3.73%	3.73%	3.73%
1b	Additional Revenue From Rate Adjustment	$30,000	$32,000	$34,000	$35,000	$37,000
1c	Service Revenue—Water Sales	843,000	888,000	936,000	985,000	1,036,000
2	Miscellaneous Fees	2,000	2,000	2,000	2,000	2,000
3	Total Operating Revenues	845,000	890,000	938,000	987,000	1,038,000
	Nonoperating Revenues					
4	Tap/Connection Fees	10,000	10,000	10,000	10,000	10,000
5	Interest Income	10,000	10,000	10,000	10,000	10,000
6	Other Revenue	5,000	5,000	5,000	5,000	5,000
7	Total Nonoperating Revenues	25,000	25,000	25,000	25,000	25,000
8	**Total Revenues**	870,000	915,000	963,000	1,012,000	1,063,000
9	**Total O&M Expenses**	459,000	473,000	537,000	555,000	570,000
10	**Subtotal: Net Operating Income**	411,000	442,000	426,000	457,000	493,000
	Debt Service					
11	Debt Service—Existing	150,000	150,000	150,000	150,000	150,000
12	Debt Service—New	0	0	9,000	87,000	101,000
13	**Total Debt Service**	150,000	150,000	159,000	237,000	251,000
14	**Rate Financed Capital Costs**	175,000	180,000	177,000	172,000	177,000
15	**NET INCOME (LOSS) FROM OPERATIONS**	86,000	112,000	90,000	48,000	65,000
16	**Plus: Beginning Cash Balance**	200,000	286,000	398,000	488,000	536,000
17	**Ending Cash Balance Before Reserves**	286,000	398,000	488,000	536,000	601,000
	RESERVES					
18	O&M Reserve	115,000	118,000	134,000	139,000	143,000
19	Capital Reserve	456,000	456,000	456,000	456,000	456,000
20	**Total Reserves**	571,000	574,000	590,000	595,000	599,000
21	**ENDING CASH BALANCE AFTER RESERVES**	($285,000)	($176,000)	($102,000)	($59,000)	$2,000
22	**CUMULATIVE REVENUE AND RESERVE DEFICIENCY (line 21 divided by line 9)**	-34%	-20%	-11%	-6%	0%
23	**DEBT SERVICE COVERAGE (line 10 divided by line 13)**	2.74	2.95	2.68	1.93	1.96

Appendix B

Examples of Financial Policies

The ability to provide high-quality water service depends not only on a reliable physical plant but also on maintaining a sound financial position. Water utilities are capital-intensive operations. Most utilities have aging infrastructure and a continual need to expand system capacity demand. Therefore, it is critical that utilities have ready access to funding for capital, whether in the bond markets, through local lending institutions, or through user revenues. The implementation of a sound financial policy is essential to protect this access.

Similar to the earlier discussion on reserve levels, many utilities, many rating agencies, and the investment community are taking critical looks at a utility's comprehensive financial policies. Besides reserve policies and levels, it is important to demonstrate policies related to cost recovery, debt management, and investment decisions. Taken together, a comprehensive policy statement will demonstrate financial strength and security.

COST RECOVERY POLICY

An important element of strong financial management practice is a cost recovery policy (CRP). In a CRP, a utility would develop a philosophy linking revenues to expenditures. A CRP should set financial goals and objectives for the utility. These goals and objectives could include securing favorable bond ratings, promoting financial stability, setting fund balance targets, and establishing a financial plan and schedule of future financial plan updates. The utility would determine a funds management plan and set rates, fees, and charges to meet those goals and objectives. Examples to be considered include:

1. *Capital projects related to growth.* Utilities should consider having these costs paid by the customers that benefit from the facilities intended to serve such growth. This can be done through system development charges, connection fees, and new account fees. If debt funding is used to finance growth-related costs, the CRP should require this debt service to be paid from system development charges or establish a process for a system development charge credit for growth expenses paid through rate revenues, depending on anticipated development activity.

2. *Renewal and replacement capital costs.* The CRP should set goals for funding renewal and replacement projects by establishing guidelines for what projects will be cash-funded and what projects will be eligible for debt funding. Typically, the guidelines will consider the cost and useful life of the capital project.

3. *Operating costs.* The utility should allocate operating costs equitably among its customer classes. The CRP should establish which operating expenses would be recovered by the fixed portion of the customer bill, which expenditures would be recovered by the volumetric portion of the bill, and which expenditures would be recovered by ancillary fees and charges.

DEBT POLICY

The environment for debt financing has changed since the credit crisis of 2008. Utilities that wish to secure debt financing should be aware that the bond market is applying due diligence criteria more stringently than in the past. A comprehensive debt policy would demonstrate to the bond market that a utility has planned for future debt, which in turn may help ensure lower financing costs.

Along with reserve levels, one of the most closely monitored indicators of utility performance is the debt service coverage ratio. This metric relates to the ability of a utility not only to meet annual principal and interest requirements but also to provide for an additional amount of utility revenues over and above debt service requirements. This additional amount provides an added level of assurance to bondholders that a utility would be able to meet all of its obligations: operational, capital, debt, and contingency. As the coverage level increases, the greater the comfort level of current and prospective investors.

It is important for a utility to monitor financial metrics annually and match to policies and required targets. Furthermore, periodic reviews of these policies would ensure a utility is meeting its current and anticipated obligations.

INVESTMENT POLICY

A utility's investment policy should consider statements that allow the utility to invest public funds in a manner that will provide the highest investment return with the maximum security while meeting daily cash flow needs. A sound investment policy should comply with applicable laws, regulations, and debt financing covenants regarding persons responsible for fund investment, how funds will be invested, and when and how a periodic review of investments will be made.

The objectives of a utility's investment activity should consider at a minimum the following features:

- Safety (protection of investment principal)
- Risk, including credit risk, interest rate risk, and custodial credit risk
- Liquidity
- Yield
- Diversification of investments
- Authorized investments

Over the years, AWWA's Board of Directors has adopted and reaffirmed its policy regarding utility finance, accounting, and rates. This policy follows as a guideline for utility development of financial policies.

AWWA POLICY—FINANCE, ACCOUNTING, AND RATES*

To this end, AWWA recognizes the following principles that water utilities should establish. Implementation of these principles can be balanced against other policy objectives; however, no policies should be adopted that compromise the long-term financial integrity of water utilities or their ability to provide service to customers. Basic financing and rate principles include:

1. Water utilities' revenues from water service charges, user rates, and capital charges (e.g., impact fees and system development charges) should be sufficient to enable utilities to provide for:
    - Annual operation and maintenance expenses;
    - Capital costs (e.g., debt service and other capital outlays); and
    - Adequate working capital and required reserves.

2. Water utilities should account for and maintain their funds in separate accounts from other governmental or owning entity operations. Water utility funds should not be diverted to uses unrelated to water utility services. Reasonable taxes, payments in lieu of taxes, and/or payments for services rendered to the water utility by a local government or other divisions of the owning entity may be included in the water utility's revenue requirements after taking into account the contribution for fire protection and other services furnished by the utility to the local government or to other divisions of the owning entity.

3. Water utilities should adopt a uniform system of accounts based on generally accepted accounting principles. Utility practices should generally follow the accounting procedures outlined in the *Financial Management for Water Utilities: Principles of Finance, Accounting, and Management Controls* published by AWWA. Modifications may be made to satisfy the financial and management control reporting needs of the utility and to meet the requirements of legislative, judicial, or regulatory bodies.

4. Water rate schedules should be designed to distribute the cost of water service equitably among each type and class of service. Non-cost of service rate-setting practices may be appropriate in some situations, subject to legal review and

* Adopted by the Board of Directors Jan. 25, 1965, revised Jan. 31, 1982, reaffirmed Jan. 25, 1987, revised Jan. 26, 1992, June 21, 1998, Jan. 16, 2005, and revised Jan. 17, 2010.

approval, provided they reflect market conditions, the benefits received by the users of the service, and an appropriate balance of the goals and objectives essen tial to the public good. Any non-cost of service rate-setting practice implemented by a utility should be fully disclosed to its customers, regulators, and the financial community. Such disclosure should identify each non-cost of service rate-setting practice, its expected benefit, and its impact on the utility's customers.

5. Water utilities should maintain asset records that detail sufficient information to provide for the monitoring and management of the physical condition of infrastructure. These asset records should also support planned and preventive maintenance programs and budgets adequate to maintain the utility's assets at a level of service consistent with good utility practice. Utilities should annually provide comparative information to customers, the financial community, and the general public about the utility's sustained capability to provide water service and generate revenue levels necessary to protect the financial investment of others in the utility. Such information could include historical expenditures for renewal and replacement during each of the past several years, as well as the revenues that would be generated under planned and adopted rates to support renewal and replacement during each of the next several years.

Appendix C

Frequently Used Financing Sources for Water Utilities

Small water utilities face the financial challenge of owning and operating a capital-intensive business. Access to capital, however, through the sale of revenue bonds and other debt instruments may not be an option. Where is another place to find money for capital improvements? The capital network in Figure C-1 shows the possible funding alternatives available from different levels of government. Local utilities gain access to capital from federal and state agencies and their nonprofit cohorts. The mutual benefit of small water utilities using the capital network is understandable.

Capital is available using a variety of loans, grants, and partnering opportunities in the network. Funds flow through the federalist system of government. Federal and state agencies pool their capital into temporary and permanent sources of funding. Funding is then distributed to qualified utilities in local government and the private sector. States are usually responsible for distributing the majority of funds. Federal agencies, however, may participate in distributing funds directly to water utilities.

Finding capital requires research. The capital network is summarized in the following lists. Sources of capital are limited to federal agencies routinely funding water utilities. Use the links to research funding criteria and identify other funding opportunities. State agencies are designed to support many activities of the federal government. A state revolving fund is a popular funding model between state and federal agencies. Use this orientation to target state research. Local water utilities take many forms of ownership. Water utilities are owned by municipalities, counties, and special districts. Investor-owned utilities may also be eligible for capital funding in some circumstances.

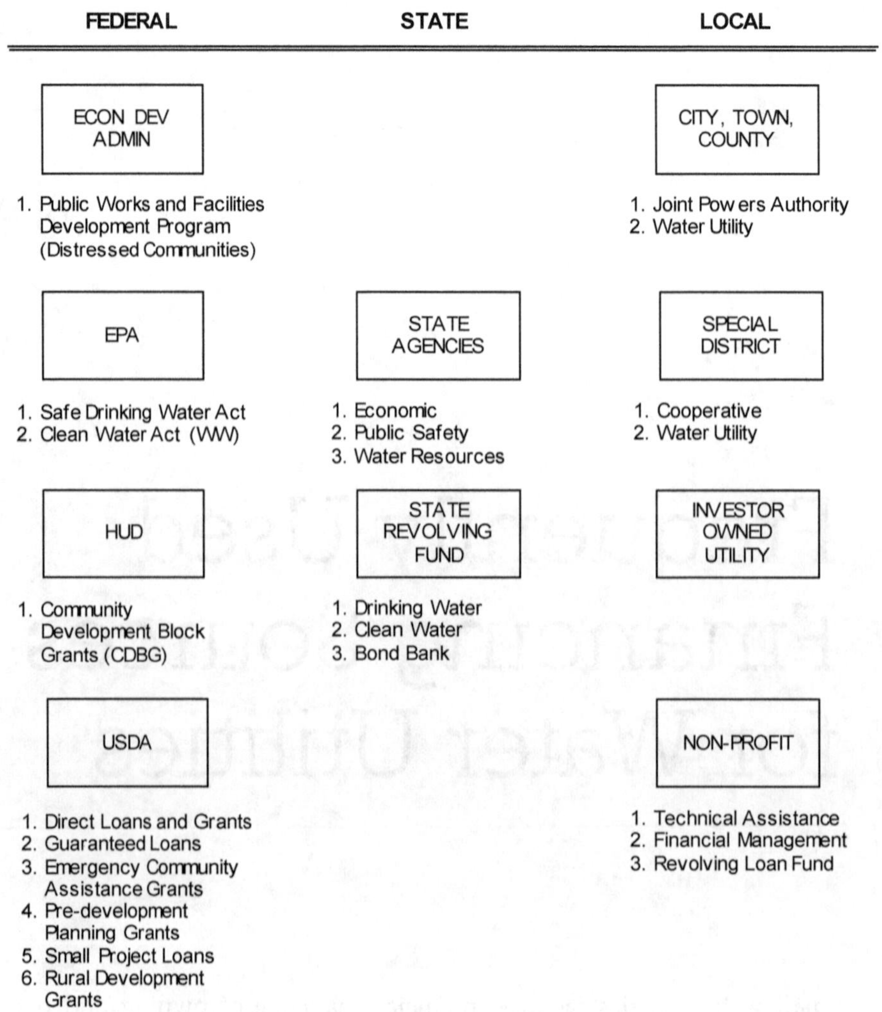

Figure C-1 Funding opportunities from loans, grants, and partnerships flow from a capital network of federal and state agencies to water utilities in local government and the private sector

Nonprofit organizations round out the capital network. They have the expertise and resources to help utilities acquire capital.

Federal

1. Economic Development Administration (EDA)
    a. Capital investments encourage overall global competitiveness and job creation based on six investment priorities. Priorities include economically distressed communities.
    b. http://www.eda.gov/

2. Environmental Protection Agency (EPA)
    a. Capital investments support the environment through water system improve ments. Funding is authorized by the federal Safe Drinking Water Act and administered by state through the Drinking Water State Revolving Fund. Capital investments are also available for wastewater through the Clean Water Act.

   b. http://water.epa.gov/grants_funding/

3. Housing and Urban Development (HUD)
   a. Capital investments facilitate community development through water system improvements. Funding is available through Community Development Block Grants.
   b. http://portal.hud.gov/hudportal/HUD?src=/program_offices/comm_planning/communitydevelopment

4. United States Department of Agriculture (USDA)
   a. Capital investments fund drinking water facilities in rural areas, cities, and towns with a population of 10,000 or less. Five funding vehicles are offered by Water and Environmental Programs.
   b. http://www.rurdev.usda.gov/UWEP_HomePage.html

State

1. State Agencies. State agencies not only contribute capital but also administer the distribution of funds for federal agencies. State and federal agencies, consequently, have common interests such as economic and community development, public safety, and water resources.

2. State Revolving Fund. State and federal agencies pool their capital to create sustainable sources of capital funding. The federal Drinking Water State Revolving Fund is a good example of this partnership. States may also establish their own capital programs.

3. Bond Bank. State agencies capitalize bond banks to offer loans at attractive interest rates and fewer transaction fees. Planning and design grants may also be available.

Nonprofit Organizations (NPOs)

1. These organizations come in many shapes and sizes. The Rural Community Assistance Partnership (RCAP) is one example. It is funded solely by federal agencies. RCAP not only provides technical expertise but also administers a lending program.

2. http://www.rcap.org/programs

The search for capital is intense. The best strategy is to be prepared. The first step is to communicate. Develop a financial plan acceptable to all interested parties. Prove annual operating and capital requirements will be fulfilled using internally generated revenue and external sources of capital. Review industry publications for guidance. Use publications from the American Water Works Association (e.g., Manuals M1 and M29) and other organizations.

The second step is to know where to look. Mine the capital network. Search for capital matching the short- and long-term needs of the water utility. Know the selection criteria, funding priorities, and schedule of federal and state agencies. Customize applications by highlighting information from the financial plan.

The third step is to ask for help. Ask for technical assistance from the cohort of non-profit organizations. Contact elected officials representing your service area. Enlist their support at every level of government.

Index

Note: *f.* indicates figure and *t.* indicates table

F

G

H

I

J

M

N

O

P

R

S

T

U

V

W

This page intentionally blank.

AWWA Manuals

M1, *Principles of Water Rates, Fees, and Charges*, #30001
M2, *Instrumentation and Control*, #30002
M3, *Safety Management for Water Utilities*, #30003
M4, *Water Fluoridation Principles and Practices*, #30004
M5, *Water Utility Management*, #30005
M6, *Water Meters—Selection, Installation, Testing, and Maintenance*, #30006
M7, *Problem Organisms in Water: Identification and Treatment*, #30007
M9, *Concrete Pressure Pipe*, #30009
M11, *Steel Pipe—A Guide for Design and Installation*, #30011
M12, *Simplified Procedures for Water Examination*, #30012
M14, *Backflow Prevention and Cross-Connection Control: Recommended Practices*, #30014
M17, *Fire Hydrants: Installation, Field Testing, and Maintenance*, #30017
M19, *Emergency Planning for Water Utilities*, #30019
M20, *Water Chlorination/Chloramination Practices and Principles*, #30020
M21, *Groundwater*, #30021
M22, *Sizing Water Service Lines and Meters*, #30022
M23, *PVC Pipe—Design and Installation*, #30023
M24, *Planning for the Distribution of Reclaimed Water*, #30024
M25, *Flexible-Membrane Covers and Linings for Potable-Water Reservoirs*, #30025
M27, *External Corrosion Control for Infrastructure Sustainability*, #30027
M28, *Rehabilitation of Water Mains*, #30028
M29, *Water Utility Capital Financing*, #30029
M30, *Precoat Filtration*, #30030
M31, *Distribution System Requirements for Fire Protection*, #30031
M32, *Computer Modeling of Water Distribution Systems*, #30032
M33, *Flowmeters in Water Supply*, #30033
M36, *Water Audits and Loss Control Programs*, #30036
M37, *Operational Control of Coagulation and Filtration Processes*, #30037
M38, *Electrodialysis and Electrodialysis Reversal*, #30038
M41, *Ductile-Iron Pipe and Fittings*, #30041
M42, *Steel Water-Storage Tanks*, #30042
M44, *Distribution Valves: Selection, Installation, Field Testing, and Maintenance*, #30044
M45, *Fiberglass Pipe Design*, #30045
M46, *Reverse Osmosis and Nanofiltration*, #30046
M47, *Capital Project Delivery*, #30047
M48, *Waterborne Pathogens*, #30048
M49, *Butterfly Valves: Torque, Head Loss, and Cavitation Analysis*, #30049
M50, *Water Resources Planning*, #30050
M51, *Air Valves: Air-Release, Air/Vacuum and Combination*, #30051
M52, *Water Conservation Programs—A Planning Manual*, #30052
M53, *Microfiltration and Ultrafiltration Membranes for Drinking Water*, #30053
M54, *Developing Rates for Small Systems*, #30054
M55, *PE Pipe—Design and Installation*, #30055
M56, *Nitrification Prevention and Control in Drinking Water*, #30056
M57, *Algae: Source to Treatment*, #30057
M58, *Internal Corrosion Control in Water Distribution Systems*, #30058
M60, *Drought Preparedness and Response*, #30060
M61, *Desalination of Seawater*, #30061
M63, *Aquifer Storage and Recovery*, #30063
M65, *On-Site Generation of Hypochlorite*, #30065
M66, *Cylinder and Vane Actuators and Controls—Design and Installation*, #30066

www.ingramcontent.com/pod-product-compliance
Lightning Source LLC
LaVergne TN
LVHW082358200926
848481LV00012B/3604